CHILDBIRTH
AND THE FUTURE
OF HOMO SAPIENS

MICHEL ODENT
CHILDBIRTH
AND THE FUTURE
OF HOMO SAPIENS

pinter
&
martin

Childbirth and the Future of Homo sapiens

First published in Great Britain by Pinter & Martin Ltd 2013

© 2013 Michel Odent

Michel Odent has asserted his moral right to be identified as the author of this work in accordance with the Copyright, Designs and Patents Act of 1988.

All rights reserved

ISBN 978-1-78066-095-0

British Library Cataloguing-in-Publication Data
A catalogue record for this book is available from the British Library.

This book is sold subject to the condition that it shall not, by way of trade and otherwise, be lent, resold, hired out, or otherwise circulated without the publisher's prior consent in any form or binding or cover other than that in which it is published and without a similar condition being imposed on the subsequent purchaser.

Set in Minion

Printed and bound in the UK by TJ International Ltd, Padstow, Cornwall

This book has been printed on paper that is sourced and harvested from sustainable forests and is FSC accredited.

Pinter & Martin Ltd
6 Effra Parade
London SW2 1PS

www.pinterandmartin.com

CONTENTS

INTRODUCTION

The long-term consequences of how we are born: an anthology of valuable studies

In the 1980s there were already theoretical reasons to claim that the way we are born might have long-term consequences.[1] What was mostly theoretical some decades ago has been confirmed by a recent accumulation of published hard data. A new generation of research is inspiring a great variety of urgent questions regarding the future of our species. This is why I find it necessary to introduce this book with an anthology of relevant articles published in authoritative medical and scientific journals. I have selected studies related to pathological conditions that are suddenly becoming more frequent in our societies and also studies whose results are particularly thought-provoking. All the references can be found at www.primalhealthresearch.com, a database specializing in studies exploring correlations between what happens during the primal period (including birth) and what happens later on.

The results of a Chinese study published in 2010 are particularly intriguing. The objective was to evaluate the likelihood of psychopathological problems in childhood in relation to how babies are born.[2] This huge study (more than 4,000 first babies born at term) was conducted in south-east China, among a population where the overall rate of caesarean sections was 56%. Three groups of children were assessed. Children in the first group were born by the vaginal route without forceps (or ventouse). Those in the second group were born with forceps (or ventouse). Those in the third group were born by programmed pre-labour c-section on maternal request. Children in the third group, born by pre-labour c-section, had the lowest risk of psychopathological problems. After controlling for many associated factors, the differences remained highly significant. One of my own interpretations is that, in the context of modern China, almost all children had probably been exposed to synthetic oxytocin, except those born by pre-labour c-section.

Several studies of autism from a primal health research perspective have a great scientific value. For example, there has been a study involving every Swedish baby born over a period of twenty years. Interestingly, whatever the country and whatever the research protocols, all these studies have detected risk factors for autism during the period surrounding birth.[3,4,5,6] In the case of a disorder that is not purely genetic, it is essential to obtain indications about the critical period for gene-environment interaction. We must emphasize that the period surrounding birth is probably also critical in the genesis of anorexia nervosa.[7,8] This fact is crucial to the hypothesis that autism and anorexia nervosa are two facets of the same disease, one variant being mostly male and the other mostly female.

A great variety of studies of asthma and allergic diseases also provide concordant results in terms of timing for gene-environment interaction.[9,10,11,12] Furthermore, the results of some studies suggest new interpretations of previously published hard data. Until recently we had probably underestimated the paramount importance of the way the gut flora is established immediately after birth, when the baby is 'entering the world of microbes'. This is becoming a hot topic at a time when many human beings are not born through the bacteriologically rich vagino-perineal route and/

or are exposed to antibiotics during the period surrounding birth. A study of risk factors for asthma and allergies may be presented as a typical example.[13] Let us also mention the results of a study suggesting that a caesarean birth is a risk factor for obesity in childhood:[14] one plausible interpretation immediately comes to mind when we remember that obese adults have altered gut floras.[15]

We must emphasise that the prevalence of all the pathological conditions we have mentioned in this short anthology has dramatically increased in recent decades. This is also the case with many so-called auto-immune diseases, when a dysregulated immune system identifies some parts of the body as 'foreign' and attacks them. A typical example is type-1 diabetes, in which antibodies destroy the pancreas. This kind of diabetes, often diagnosed early in life, is fatal when insulin treatment is not available. An accumulation of data suggests that the period surrounding birth is crucial in its genesis.[16,17]

In general, it is difficult to obtain valuable studies about behavioural problems that are at the frontier between personality traits and pathological conditions. This is why thought-provoking studies about juvenile criminality,[18] suicides[19,20,21] and drug addictions[22,23] need to be replicated in order to moderate the possible effects of a great variety of confounding factors.

All those interested in the future of humanity, whatever their background, need to be aware of this new generation of research. We must, however, deliver a warning for those who are not familiar with statistical language. Epidemiologists provide conclusions in terms of tendencies, risk factors and statistically significant differences. They need huge numbers to make their studies valuable, but they cannot say anything about individuals and particular cases. This is why readers of this book must try to forget their own family and friends.

One of the functions of the Primal Health Research Database is to introduce the necessary collective dimension. Our topic is the future of humanity.

CHAPTER ONE

Ecce Homo

Our first objective is to raise a question: should we expect transformations of our species in relation to the way babies are born?

All aspects of human lifestyle have been deeply modified in recent decades. This indisputable fact has inspired both comments about recent detectable transformations of *Homo sapiens*, as well as questions about the future of our species. It is noticeable that neither in academic circles nor among the media is the period surrounding birth usually taken into consideration, although it is undoubtedly a phase of human life that has been radically turned upside-down. Several scientific disciplines now claim that it is a critical period in the formation of individuals.

Before considering the future, we'll take as a point of departure a presentation of *Homo sapiens*. How can we summarize our understanding of human nature?

The capacity to think has traditionally been considered the main characteristic of our species. It is significant that the English-Dutch word 'man' (German: 'mann', 'mensch'; Danish: 'mand') probably comes from a Sanskrit term that means 'thinking'. According to Blaise Pascal, man is '*un roseau pensant*' (a thinking reed). In the current scientific context, a common presentation of *Homo sapiens* is not radically different from the traditional one, although it is expressed in different language. Today we can present ourselves as members of the chimpanzee family with a gigantic brain of enormous complexity. We have developed to an extreme degree the part of the brain called the neocortex. This is how we interpret our mental capability, which includes the capacity for abstract reasoning, language, introspection, problem-solving and use of tools.

It is the objective of a great variety of scientific disciplines to improve our understanding of human nature. The extreme degree of specialization of modern scientists is becoming an obstacle to presenting a synthetic overview of the particularities of *Homo sapiens*. It makes us think of the famous story of a group of blind men who were asked to determine what an elephant looked like by feeling different parts of its body. The blind man who felt a leg said the elephant is like a pillar; the one who felt the tail said the elephant is like a rope; the one who felt the trunk said the elephant is like the branch of a tree; the one who felt the ear said the elephant is like a hand-held fan; the one who felt the belly said the elephant is like a wall; and the one who felt the tusk said the elephant is like a solid pipe. This parable is more significant than ever, at a time when there is an urgent need for communication and respect for different perspectives.

Bacteriology – more precisely molecular microbiology – is a typical example of a discipline advancing at such a high speed that *Homo sapiens* can now be visualized in the framework of the 'microbiome revolution'. A human being may be considered an ecosystem with a constant interaction between the hundreds of trillions of micro-organisms that colonise the body (the 'microbiome') and the trillions of cells that are the products of our genes. In other words, it appears today that our health and our behaviour are highly influenced by our gut flora and our skin flora. The point is that each person has a relatively different microbiome. Our microbiome – as a part of our personality – is to a certain extent established at birth, according to the first microbes that colonize the newborn's body.

Advances in our understanding of the specific nutritional needs of the brain led to the concept of 'brain selective nutrients'. This concept has important implications regarding a species characterized by an exceptionally high degree of encephalization. Iodine is a typical brain selective nutrient, since it is necessary for the production of thyroid hormones that are involved in brain energy metabolism. Iodine deficiency is associated with impaired brain development and function. Yet there is no mechanism to decrease iodine excretion in the urine and therefore to store it.[1] This physiological perspective suggests that *Homo sapiens* is adapted to an environment continuously providing a sufficient amount of iodine. In practice this implies access to the seafood chain. It is significant that iodine deficiency is the most common nutritional deficiency, affecting nearly two billion people globally,[2] despite the fact that iodine is the only nutrient for which many governments impose supplementation by law.

While iodine is considered the primary brain selective mineral, docosahexaenoic acid (DHA) is considered the brain selective fatty acid.[3] The molecule of DHA is as long as possible (twenty-two carbons) and as unsaturated as possible (six double bonds). It belongs to the omega-3 family. This fatty acid is preformed and abundant only in the seafood chain. Interestingly, humans have a low capacity to make DHA. The association of a large brain with a weak enzymatic system of desaturation–elongation suggests that human beings need to have access to the seafood chain in order to develop their full potential.

From a nutritional perspective, *Homo sapiens* appears to be an ape adapted to the coast. Today this idea should lead us to attach a renewed importance to what has been called in the past 'the aquatic ape hypothesis of the origin of man'.

Apart from brain size there are dozens of other features which make us different from our very close relative, the common chimpanzee: nakedness, a layer of fat attached to the skin, the general shape of our body (with the hind limbs forming an extension of the trunk), a comparatively low basal body temperature, the development of a prominent nose, large empty sinuses on each side of the nasal cavities, a low larynx, a reduced number of red blood cells, anatomical particularities of the hands and feet, and a layer of vernix caseosa covering the skin of the newborn baby being among the main differences. All these features are shared with mammals adapted to the sea and are suggestive of adaptation to a coastal environment.

This new vision of *Homo sapiens* was first proposed independently by Max Westenhofer in Berlin (1942) and by Alister Hardy in Oxford (1960), but it is the British science writer Elaine Morgan who has championed the cause in her books[4,5,6] and in the seminars she has organized in order to constantly strengthen the theory. This new theoretical framework has been recently updated through a collective academic book[7] and a London conference on human evolution that I participated in.[8]

After this brief twenty-first century overview of our understanding of human nature, we are in a position to phrase appropriate questions about the future of our species.

CHAPTER TWO

Evolution revisited

Until the development of emerging disciplines such as epigenetics, it was provocative to focus on indisputable facts such as the inheritance of acquired characteristics. It was also provocative to focus on the weakening of organs and physiological functions that are not used or are disused. There was an obvious reason: during a transitory period that started in the middle of the twentieth century, the dominant paradigm – often called Neo-Darwinism – was limited to the effect of the integration of genetics and Darwinian evolution, after what had been called 'the modern synthesis'.[1] Genetic mutation was considered the ultimate source of variation within populations. Natural selection was considered the main evolutionary force which could produce adaptation. Evolution of the species was understood as a very slow process, measured by mutation rates. Within this restrictive theoretical framework, there was a lack of curiosity about the possible fast and spectacular transformations of species. The recent development of epigenetics has heralded a new phase in our understanding of evolution and an unprecedented interest in the probable transformations of the genus *Homo* in the near future.

Speedy transformations under the effects of environmental factors are well documented among mammals in general and the genus *Homo* in particular. For example, it does not take a long time for the process of domestication of mammals to modify brain structures and behaviours. Domesticated animals have few opportunities to take the initiative, to struggle for life and to compete. Compared with wild animals, they have few opportunities to exercise their brains. In a great variety of mammals, such as pigs, sheep, dogs, cats, camels, ferrets and mink, one of the effects of domestication is a significant reduction in brain size.[2] The changes in the brain that transform a wild creature into a highly domesticated strain happen very rapidly in terms of evolution – after only 120 years of domestication, a brain size reduction of about 20% has been observed in mink.[3]

Significant transformations among humans are also well documented. The recent history of Korea offers a typical example. After the Second World War, Korea was artificially divided into a Northern part (above the 38th parallel) and a Southern part. North and South Koreans share genetic ancestry and, until the middle of the twentieth century, the same lifestyle and the same morphological traits. Today South Koreans are on average 12cm taller than their North Korean counterparts.

All these rapid transformations inside the genus *Homo* cannot be explained by changes in DNA sequences: perspectives other than 'evolutionary genetics' are needed. This is why the advent of epigenetics is a crucial step in the history of biological sciences and in our interpretation of important aspects of the process of evolution, which is not limited to the advent of new species.

This emerging discipline is based on the concept of gene expression. Some genes may be allocated a kind of label (an 'epigenetic marker') that makes them silent. This marker can be a DNA methylation; it can also be a change in the nuclear protein content, such as modifications to histones. The phenomenon of gene expression is influenced by environmental factors, particularly during what we call 'the primal period' (foetal life, birth, and the year following birth). From an overview of the Primal Health Research database (epidemiological studies exploring correlations between what happens during the primal period and what happens later on),[4] it appears that the nature of an environmental factor is often less important than the time of exposure to this factor. Our database has become a unique tool that provides some clues about the critical periods affecting the genesis of diseases and personality traits. From a practical perspective it is often more important to identify such critical periods of gene-environment interaction than to identify the actual genes involved or evaluate the comparative roles of genetic and environmental factors.

In the age of epigenetics, we have been encouraged to pay renewed attention to studies exploring the trans-generational effects of what happens during the primal period. It now appears that epigenetic markers (the 'epigenome') may be, to a certain extent, transmitted to subsequent generations. Understanding one of the mechanisms through which acquired traits can be transmitted to subsequent generations is an important factor in our understanding of the transformation of species and their adaptation to environmental factors. While during the decades preceding the development of epigenetics the focus had been entirely on Neo-Darwinian slow mechanisms of evolution, today there is a sudden renewed interest in Lamarck's theories of 'use and disuse' and the 'inheritance of acquired characteristics'. Darwin himself was born in the year when Lamarck was publishing *Philosophie zoologique* and was openly influenced by Lamarckian theories. Interestingly, Darwin also wrote a book entitled *The Variation of Animals and Plants Under Domestication*. In the language of the twenty-first century, we might say that the name of Lamarck is associated with 'soft inheritance' (epigenetic information), while the name of Darwin is associated with 'hard inheritance' (genetic information).[5]

In the current scientific context, it appears that the mother transmits more than her genome to her offspring. She also transmits some epigenetic markers, and, to a certain extent, the microorganisms that colonize her body, her microbiome.

Today, at a time when human lifestyles are changing so quickly, we live long enough to watch transformations of the genus *Homo* happening before our eyes.

After these comments on evolution, we are in a position to wonder what changes to *Homo sapiens* we can expect in the near future, in relation to how babies are born.

CHAPTER THREE

The future of the human oxytocin system

In the age of epigenetics, at a time when there is a renewed interest in theories of evolution that preceded Darwin, some facts become acceptable since they are open to interpretation. In particular it is becoming acceptable to claim that when organs or physiological functions are underused they become weaker from generation to generation. There is one spectacular example of a human physiological function that has become suddenly less useful. It is the oxytocin system. The paramount importance of high peaks of oxytocin in childbirth has been well known for more than a century (oxytocin literally means 'fast birth').

An underused physiological system

Until recently, in order to have a baby and to deliver the placenta, a woman was obliged to rely on the release of her natural oxytocin, as the main component of a complex hormonal flow. Today, the use of synthetic oxytocin ('syntocinon' or 'Pitocin') is by far the most common medical intervention in childbirth. Most women are under the effects of synthetic oxytocin when giving birth to both the baby and the placenta. In the case of a caesarean during labour, a drip of synthetic oxytocin has usually been used before the decision to operate was taken. The rates of labour induction are high: in practice, labour induction is almost always associated with the use of synthetic oxytocin. Furthermore, synthetic oxytocin is commonly used as a uterotonic agent during a caesarean section and, of course, a woman does not release her natural hormones when giving birth by pre-labour caesarean section. We can observe that, at a planetary level, the number of women who give birth to babies and placentas thanks to the activity of their own oxytocin system is becoming insignificant.

The oxytocin system is also underused to feed babies. Let us recall that a milk ejection reflex is induced by oxytocin release. When considering the average number of babies per woman in modern societies and the very short average duration of breastfeeding, it is easy to reckon that the number of milk ejection reflexes in the life of a modern woman is very small compared with what it has been in other societies.

There are no other examples of physiological systems that have suddenly been made useless under the effects of changes in lifestyle. I use the term 'system' to emphasize that we are referring to the capacity to secrete oxytocin, to use it as a neuromodulator, to store it in the posterior pituitary gland, to release it in a pulsatile effective way, and to develop receptors. We therefore have reasons to anticipate a weakening of this physiological system. This may have far-reaching consequences, since oxytocin is involved in all aspects of our reproductive/sexual life, in socialization, and in all facets of the capacity to love, which might include respect for 'Mother Earth'.

I dare to go one step further and suggest that perhaps this physiological system is already deteriorating. We can express concern about this possibility by bringing together a great diversity of published data relating to childbirth, breastfeeding, genital sexuality and capacity for empathy.

The capacity to give birth

A synthesis of recently published statistics suggests the increased difficulties of modern births. Other factors than an inappropriate hospital environment must be taken into account, since these increased difficulties are particularly significant when considering the case of home births. The difficulties of home births have been evaluated by an authoritative meta-analysis (combining data provided by several studies using similar protocols) published in *The American Journal of Obstetrics and Gynecology*.[1] All the studies in this article had compared the outcomes of planned home births with planned hospital births, which is the best method since randomized controlled trials are not feasible. Let us recall that studies comparing outcomes of actual home with actual hospital births underestimate the risks associated with planned home births, because they do not take into account the transfers to hospital during labour. The studies included in the meta-analysis were conducted in developed western countries (USA, UK, Canada, Australia, Switzerland, Netherlands and Sweden) and published in the English language in peer-reviewed medical journals.

Among the results provided by this meta-analysis we must keep in mind one significant piece of data. It is noticeable that in the planned home birth group up to 37% of women giving birth to their first baby required transfer to a hospital during labour. It is also noticeable that the neonatal death rates (deaths between birth and the age of twenty-eight days) appear twice as high after planned home births than after planned hospital births, and almost three times as high among newborn babies without abnormalities.

The data provided by a recent study covering the whole region of Utrecht, Holland, are particularly thought-provoking because they might lead us to reconsider the very special Dutch system. This system is different from all other obstetric care systems because it is based on a well-defined distribution between primary and secondary care.

In practice women with a low risk of pathology are in the hands of midwives, while the others are in the hands of obstetricians. One of the effects of this classification is a rate of home births above 20%. According to this study, the risks of perinatal death were significantly increased when the pregnancy was classified as low risk.[2] Paradoxically, the rates of transfers to a paediatric unit are not higher after a high-risk pregnancy. A study by the Birthplace in England Collaborative Group reached similar conclusions: transfers from non-obstetric unit settings were in the region of 36% to 45% among women giving birth to their first baby.[3] The difficulties of modern births in general are confirmed by a Chinese study in a population where the overall rate of caesarean section was 56%. In this study of the risks of psychopathological problems in relation to how babies are born, researchers could include a group born with forceps or ventouse.[4] This fact suggests that a rate of 56% caesarean section was not sufficient to prevent difficult births by the vaginal routes. Of course many other factors than a declining capacity to give birth must be taken into account to interpret such statistics.

Having been involved in childbirth for more than half a century (I spent six months in the maternity unit of a Paris hospital in 1953), I realize the importance of an American study about 'Changes in labor patterns over 50 years'.[5] The authors compared a first group of nearly 40,000 births that occurred between 1959 and 1966 and a second group of nearly 100,000 births that occurred between 2002 and 2008. They only looked at births of one baby at term, with head presentation and spontaneous initiation of labour. After taking into account many factors such as the age, the height and the weight of the mother, it appeared that the duration of the first stage of labour was dramatically longer in the second group. It was two and a half hours longer in the case of a first baby, and two hours in the other cases. For practitioners of my generation this study is demonstrating the obvious.

The capacity to breastfeed

In spite of intense public health campaigns to promote breastfeeding, and although contemporary populations are better informed than ever about the irreplaceable value of mother's milk, recent breastfeeding statistics are worrying. For example, among the fourteen countries that belong to the OECD (Organisation for Economic Co-Operation and Development), the rates of exclusive breastfeeding at six months are in the region of 25%.[6] In such developed countries these statistical results cannot be explained by a lack of knowledge of the value of exclusive breastfeeding during the first six months. It is commonplace to provide many complementary interpretations. One of them is that we are still in the aftermath of a period when breastfeeding was devalued. In Western Europe and North America this period started soon after the Second World War with the advent of 'humanized milk' (artificial milk, or infant formula) and lasted until the 1970s. Another plausible interpretation is based on the strong connections between birth physiology and the physiology of lactation. In the age of synthetic oxytocin and simplified caesarean technique, when most women don't give birth by themselves, it would be surprising to have good breastfeeding statistics. It might simply be that the capacity to breastfeed is declining. This is a point of view expressed by breastfeeding advisers and lactation consultants who have many decades of practice.

Genital sexuality

Any question about the evolution of the human oxytocin system leads us to look also at genital sexuality. Erection and vaginal lubrication are associated with oxytocin release. There is no sexual intercourse without activation of the oxytocin system. For obvious reasons, it is difficult to provide valuable statistics about the incidence of sexual dysfunction. However, there is such an accumulation of significant data that *The Journal of the American Medical Association* (*JAMA*) could classify sexual dysfunction among the new public health preoccupations.[7] One can observe that most sex-therapists are overworked. One can also observe that drugs to correct sexual dysfunction top the lists of pharmaceutical substances in terms of commercial value. Once more, many interpretations can be suggested, but one cannot dismiss the hypothesis of a declining oxytocin system.

Capacity for empathy

It is precisely because statistics about the different episodes of human reproductive life can inspire multiple interpretations that we must also take into account our current knowledge of the evolution of the capacity to love. Modern psychological sciences have developed scores to evaluate personality traits such as the capacity for empathy. At the annual meeting of the Association for Psychological Science, in June 2010, a synthesis of seventy-two studies of the evolution of personality traits of American students (college graduates) between 1979 and 2009 was presented.[8] According to this research, college graduates are 40% less empathetic than those of two or three decades ago. The decline has been progressive, particularly after the year 2000. Once more, this is compatible with a declining oxytocin system.

Should we learn from bulldogs?

By combining different ways of looking at the situation, we have raised questions that justify further research. Animal experiments should be feasible with non-human mammals. Rats, for example, develop rapidly in infancy and become sexually mature at the age of six weeks (more than 100 times faster than humans). In a year the number of generations of rats is roughly the same as in two centuries of humans. It would not take long to observe the characteristics of rats after a dozen caesarean-born generations.

Without waiting for the results of such animal experiments, we might wonder what we can learn from breeds of dog that have had medically assisted births for many decades and are now almost routinely born by caesarean section. Today the rates of caesarean section among English bulldogs are in the region of 90% and artificial insemination is common practice. In other words, taking into account the age at puberty, it should be feasible to study the oxytocin system of bulldogs that already have twenty generations of caesarean births (and artificial inseminations) behind them, the equivalent of several centuries among humans. Comparisons are still possible with other members of the same breed born via the vaginal route.

Today the life of bulldogs – particularly their reproductive life – is almost entirely in the hands of veterinary medicine. There are real questions to be asked, and answered, here.

CHAPTER FOUR

A landmark in the evolution of brain size?

Starting from the observation that the human oxytocin system is now underused, particularly in the critical period surrounding birth, we suggested possible transformations of *Homo sapiens* from a physiological perspective. We must add that in the age of simplified fast techniques of caesarean section, morphologic transformations should also be expected, particularly in terms of head circumference.[1] Are we reaching a landmark in the evolution of brain size?

Inextensible limits

Until recently it was commonly accepted that, for obstetrical reasons, the development of the human brain had reached its limits. At term, the smaller diameter of the baby's head (which is not exactly a sphere) is roughly the same as the larger diameter of the mother's pelvis (which is not exactly a cone). The evolutionary process adopted a combination of solutions in order to reach the limits of what is possible.

The first solution was to make pregnancy as short as possible, so that, in a sense, the human baby is born prematurely. It is as if there is a 'primal period'[2] that includes a phase of 'internal gestation' followed by a phase of 'external gestation' in a social milieu. The comparative immaturity of the human newborn baby cannot be dissociated from the development of strong social groups. I have suggested that an early birth might have multiple advantages in terms of brain development: the extra-uterine world can provide a much greater variety of complex personalized sensory stimulation than the prenatal environment.[3] We have realized recently that the pregnant mother can, to a certain extent, adapt the size of the foetus to her own size by modulating the blood flow and the availability of nutrients to the foetus. That is why small surrogate mothers carrying donor embryos from much larger genetic parents give birth to smaller babies than might have been anticipated.

From a mechanical point of view, the baby's head must be as flexed as possible, so that the smaller diameter is presenting itself, before spiralling down to get out of the maternal pelvis. The birth of humans is a complex asymmetrical phenomenon, the maternal pelvis being widest transversally at the entrance and widest longitudinally at the exit. A process of 'moulding' can slightly reshape the baby's skull if necessary.

When mentioning the mechanical particularities of human birth, one cannot help referring to and comparing ourselves with our close relatives, chimpanzees. The head of a baby chimpanzee at term occupies a significantly smaller space in the maternal pelvis, and the vulva of the mother is perfectly centred, so that the descent of the baby's head is as symmetrical and as direct as possible. It seems that since we separated from the other chimpanzees, and throughout the evolution of the hominid species, there has been a conflict between moving upright on two feet and at the same time a tendency towards a larger and larger brain. The brain of the modern *Homo* is three times bigger than the brain of our famous ancestor Lucy. There is a conflict in our species because the pelvis adapted to the upright posture must be narrow enough to allow the legs to be close together under the spine, which facilitates the transfer of forces from legs to spine when running. An upright posture is a prerequisite for brain development. We can carry heavy weights on our head when we are upright. Mammals walking on all fours cannot do the same. That is apparently why the process of evolution found other solutions than an enlarged female pelvis in order to make the birth of the 'big-brained ape' possible: the faster our ancestors could run, the more likely they were to survive.

There is another reason why one can claim that the development of brain size has reached its limits. When we refer to brain development, we are not precise enough. We should emphasise that it is not the whole brain that is dramatically developed in our species. It is that part of the brain called the neocortex. The neocortex can be presented as a sort of super computer that was originally at the service of the vital archaic brain structures. The point is that this 'new brain' tends to wield most power and it can inhibit the activity of the primitive one, particularly during an involuntary process such as birth. However, some modern women are still able to release the necessary hormones and give birth with their own resources, on condition that the activity of their powerful neocortex is dramatically reduced. From this point of view it appears once more that the development of the human brain has reached its limits. A more developed neocortex might make the birth process impossible.

Pulverized limits

All these apparently fixed limits to brain evolution are swept away with the advent of the safe caesarean. Until now babies who had too large a head to pass through the maternal pelvic opening did not survive birth and therefore could not transmit this tendency for larger brain size on to future generations. Today continued evolution of brain size is possible once more.

We, as members of the species *Homo Sapiens*, are to a great extent different from the other primates, particularly our close cousins the chimpanzees, in terms of fat metabolism.

We are characterized by a huge capacity to transport fatty molecules and fatty particles to certain parts of the body, such as the skin and the brain. Only humans have a layer of fat attached to the skin and an ability to accumulate lipids in such places as breasts and buttocks for females, and abdomen for males. Modern lifestyle is associated with a tendency to make the layer of fat beneath the skin thicker and thicker.

Humans are first and foremost characterized by a unique capacity to transport to the brain specific molecules of fatty acids that are necessary for its development. The developing brain has a real thirst for the fatty acid commonly called DHA: fifty percent of the molecules of fatty acids that are incorporated into the developing brain are represented by DHA. This very long-chain polyunsaturated molecule of the omega-3 family is preformed and abundant only in the seafood chain. It is justified to claim that the transmission of the tendency for larger brains is suddenly made possible in the age of the safe caesarean, particularly in populations where preformed DHA is an abundant component of the diet. This is suggested by the results of a study we started in 1991 in a London hospital (Whipps Cross). The objective was to evaluate the possible effects in the perinatal period of simply encouraging pregnant women to consume sea fish.[4] Four hundred and ninety-nine pregnant women attending selected clinics for antenatal care before twenty weeks of gestation were offered twenty minutes of nutritional advice. For each woman interviewed a corresponding control was established. There was one highly significant difference between the two groups in the perinatal period: the mean neonatal head circumference was greater in the study group (34.65 cm v 34.45 cm 95% CI 0.01-0.39). Our Whipps Cross study was replicated and enlarged at Wolverhampton New Cross Hospital.[5] Again, the most significant difference was related to head circumference at birth. Among the 1,607 cases in the study group, the mean head circumference was 34.54 cm, against 34.32 cm among the 1,078 cases in the control group (95% CI 0.10-0.35, $p<0.001$). The statistical significance remained the same after adjustment for gestational age and sex.

It is noticeable that the vision of the caesarean as a landmark in the evolution of brain size was not originally that of an expert in evolutionary anthropology or childbirth. Jane English was born by 'non-labour' classical caesarean in 1942. When she became an adult, around the age of thirty, she started to use 'caesarean-born' as a 'lens through which to look at the world and at herself'. This is how she realized that transformations of *Homo sapiens* can be induced by our ingeniousness. Her interest in the implications of being caesarean-born was the reason for *Different Doorway*, the book she published in 1985.[6]

I was lucky enough to meet Jane English in the 1980s. Since that time I have started to understand why the history of childbirth is at a turning point, why the history of mankind is at a turning point, and even why the process of human evolution itself is at a turning point.

CHAPTER FIVE

'Microbes Maketh Man'

This headline, which appeared on the cover page of a summer 2012 issue of *The Economist*, is an exemplary illustration of the sudden perception by the media of the importance of the 'Microbiome Revolution'. 'Germs are us', the title of an article in *The New Yorker*, is another typical example.

In the current scientific context we have presented *Homo sapiens* as an ecosystem, with a constant interaction between the trillions of cells that are the products of our genes (the 'host') and the hundreds of trillions of micro-organisms that colonise the body (the 'microbiome').

One may wonder why it took so long to realize the alignment of interest of the host and the microbiome. While the host offers raw materials and shelter, hundreds of trillions of micro-organisms feed and protect their host. It is as if, from Pasteur's era until recently, there was a deep-rooted cultural conditioning associating microbes and diseases and classifying all microbes as enemies. The focus was on pathological conditions associated with an unbalanced ecosystem. This conditioning has been reinforced during the twentieth century, particularly with the advent of antibiotics that are effective at dealing with specific infections. This cultural conditioning has not been seriously challenged as long as bacteriologists could only look at microscopes and cultivate microbes on Petri dishes. Bacteriologists started to dramatically enlarge their horizons with the power of computer processing and new DNA sequencing technologies: not all bacteria seen under the microscope can be cultured, since their growth conditions are unknown. Because molecular techniques are culture-independent, bacteriologists can now see the 'unseen majority'.

It has been an important step in the history of science to realize that the body of each of us is colonized by more micro-organisms than there are human beings on planet earth. This fact inspired new questions about the multiple vital functions of our microbiome.

The main functions of our gut flora have recently been clarified. The highest number of micro-organisms is found in the large intestine. Their well-known functions include energy extraction from food, digestion of polysaccharides and synthesis of vitamin K and the B vitamin complex. The bacterial synthesis of vitamin B12 must be mentioned, although vitamin B12 brought by food is mostly absorbed in the ileum (the end of the small intestine). The protective role of the 'good' microbes against harmful microbes was understood long before the microbiome revolution by a small number of clear-sighted practitioners. As early as during the 1960s it was found, during epidemics in nurseries, that the colonization of babies with virulent staphylococci could be prevented by early voluntary contamination (nasal or umbilical) with a strain of staphylococcus selected because of its very low virulence.[1] The role of the gut flora as an 'educator' of the immune system is a comparatively new topic. The microbiome revolution is at the root of a real rethinking of the immune system. The immune system may be compared to a sensory organ that needs stimulation to develop properly. There are even studies suggesting that the education of the immune system starts during foetal life, being modulated, in particular, by the gut flora of the mother.[2]

Studies of the 'Gut-Brain connection' are among the most promising avenues for research and until now we have been learning from animal experiments. It appears that there is a critical period early in life when gut micro-organisms affect the brain and change the behaviour in later life. The most significant study is with mice that are bred to have no digestive bacteria. Scientists can then introduce the bacteria or not. Researchers found that the no-bacteria mice had much more hyperactive and risky behaviour as adults. If they were given normal bacteria early in their life then they grew up with the same normal behaviour traits as control mice. If they were given normal bacteria later in life the hyperactive/risky behaviour was already established.[3] The results of such animal experiments support the concept of 'Gut and Psychology Syndrome' (GAPS), a condition that takes for granted a connection between the functions of the digestive system and the functions of the brain.

While until now most studies have focused on the colonization of the digestive tract (including the mouth), we must also realize that trillions of bacteria, fungi, viruses, archaea, and small arthropods colonize the skin surface, collectively comprising the skin microbiome. Microbial skin colonization has been shown to critically affect the development of the skin's immune function… another vital avenue for research.[4] The importance of some parts of the skin microbiome is probably underestimated. The navel is the best example of a bacteriologically rich zone that remains relatively unexplored.[5]

As for the studies of the milk microbiome, they are still in a preliminary phase.

We cannot dissociate the current interest in the functions of the microbiome from the need to interpret the spectacular increase in the incidence of certain chronic diseases. It appears, from an exploration of the Primal Health Research Database (www.primalhealthresearch.com) that epidemiologists detect risk factors for such diseases in the period surrounding birth. This is the case, for example, of obesity, type-2 diabetes, autism, anorexia nervosa, asthma, allergic diseases, all sorts of autoimmune diseases and inflammatory bowel disease. This sudden upheaval in the comparative incidence of diseases is a source of theories and hypotheses. It makes sense to look at the critical period when the germ-free foetus is suddenly entering the world of microbes. In other words there are new reasons to study childbirth from a bacteriological perspective.

From the early days of microbiology until the 1970s, one of the roles of midwives and doctors involved in childbirth was to protect the newborn babies against all microbes, including those of maternal origin. It was usual to shave the mother at the beginning of labour, to give her an enema, and to put antiseptic solutions around the nipple. This was part of the framework of the dominant cultural conditioning, associating microbes and diseases.

A new step in the history of our understanding of childbirth from a bacteriological perspective was taken with studies demonstrating that, compared with the placenta of other mammals, the human placenta is highly effective at transferring Immunoglobulin G (IgG) to the foetus.[6] While in our species the levels of IgG of a neonate born at term are at least 100% of the maternal levels, among bovines, for example, they can be below 10%. Clearly the main preoccupations are not the same among humans as among other mammals. The newborn calf is immediately dependent on antibodies provided by early colostrum. In other words, among many species of mammals, such as bovines, the colostrum is, strictly speaking, vital. Among humans, the main preoccupation must be phrased differently: which microbes will be the first to colonize the germ-free newborn's body? A well-known concept used by bacteriologists is a reason to give great importance to this question. 'The race for the surface' means that the winners of the race to reach a germ-free territory will likely be the rulers of that territory.

It is clear today that from a bacteriological point of view a newborn baby ideally needs to be urgently in contact with only one person – her mother. It is also clear that the human mammal has been programmed to enter the world via the bacteriologically rich perineal zone: this is a sort of guarantee that the newborn baby – particularly her digestive tract and her skin – will be immediately contaminated by a great variety of friendly germs carried by her mother. It is significant that the vaginal flora changes throughout the pregnancy, culminating in a proliferation of friendly bacteria such as *Lactobacillus Johnsonii*.[7,8] It is as if the mother's body is preparing itself for the colonization of the neonate.

However, the general rule of an easy placental transfer of antibodies, particularly intense from thirty-eight weeks onwards,[9] must be modulated. One must take into account that there are four subclasses of IgG and that the transfer of the subclass 2 is not as effective as the transfer of the other subclasses.[10,11] This is a way to interpret the apparently mysterious vulnerability of human babies – particularly premature babies – to streptococci B transmitted by the mother.

We notice that for thousands of years all human groups have dramatically interfered in the process of microbial colonization of the newborn's body via beliefs and rituals. The belief that the early colostrum is harmful is a typical example. In general, mothers and newborn babies have been separated and the initiation of breastfeeding has been delayed.

Today, in the age of medicalization of childbirth, there are new obvious and powerful ways to interfere. Most babies are born in a bacteriologically unfamiliar environment. It is easy to convince anyone that babies born vaginally and babies born by caesarean enter the world of microbes in radically different ways. Furthermore, the exposure of foetuses to antibiotics in the perinatal period is common. Antibiotics are used in frequent situations, such as detection of streptococci B, premature rupture of the membranes, and also caesarean sections: some public health organizations, such as NICE (National Institute for Clinical Excellence) in the UK, officially recommend injecting antibiotics before starting a caesarean.[12]

Childbirth, from a bacteriological perspective, is becoming an area of debate now that an accumulation of data suggests that the way the newborn's body is colonised immediately after birth can have at least medium-term consequences.

Among these consequences, consider the conclusion of studies comparing the activity of cells with immune action in the blood of babies born either vaginally or by caesarean.[13] The influence of the way the baby is born on the immune response is still detectable at the age of six months.[14] Finnish studies explored the faecal flora of thirty-four children born vaginally and thirty children born by caesarean with antibiotic prophylaxis, at the ages of three to five, ten, thirty, sixty and 180 days. The faecal colonization of infants born by caesarean was delayed. The faecal flora was still disturbed at the age of six months among the caesarean-born children.[15]

Since medium-term consequences can be demonstrated, it is justified to offer plausible interpretations of the incidence of chronic diseases in the framework of the microbiome revolution. Recent studies in several fields of medicine have suggested the long-term consequences of the way the gut flora is established in the perinatal period, in relation to the mode and place of delivery. According to a Dutch study, vaginal home delivery, compared with vaginal hospital delivery, is associated with a decreased risk of eczema, sensitisation to food allergens, and asthma. Mediation analysis showed that the effects of mode and place of delivery on atopic outcomes were mediated by *C difficile* colonization.[16] The results of a breakthrough article in *Nature* identified the gut flora as a contributing factor to the pathophysiology of obesity. Microbial populations in the gut are different between obese and lean people: in mice and humans, obesity is associated with changes in the relative abundance of the two dominant bacterial divisions (the *Bacteroidetes* and the *Firmicutes*).[17] There are reasons to assume that such changes might have started at birth. A study demonstrating that a caesarean birth is a risk factor for obesity at age three must be seriously considered.[18] A new generation of studies of gut flora can establish new links between obesity and type 2 diabetes, since bacterial populations in the gut of diabetics differ from those of non-diabetics.[19]

We cannot dissociate the questions related to the microbial colonization of the digestive tract from the questions related to the colonization of the mouth. The *Journal of Dental Research* has published an authoritative study demonstrating that the mode of delivery (vaginal birth compared with c-section) affects oral microbiota in infants, and therefore dental health later on in life.[20]

In such a context, we might summarize our main reasons for concern by claiming that, in the age of simplified techniques of caesarean section, which is also the age of antibiotics, the immune system of many newborn babies is deprived of vital stimulation at a critical phase of its development.[21] Preliminary studies of the milk microbiome in relation to the way babies are born suggest that, from a bacteriological perspective, the significant differences are not between vaginal birth or caesarean, but between pre-labour and in-labour caesarean sections.[22] Such differences according to the timing of the operation are confirmed by a Canadian study of the gut flora of four-month-old babies.[23] In general, and not only in the field of bacteriology, there is a lack of research comparing the effects of pre-labour and in-labour c-sections.

Anyway, we must prepare for a fast evolution in the relationship between human beings and the world of microbes. In terms of importance in the history of biological sciences, the microbiome revolution can be compared to the advent of epigenetics.

In the same way that the advent of epigenetics has suddenly made topical the theories of Lamarck, which preceded the work of Darwin, the microbiome revolution should save from oblivion the work of Antoine Béchamp, who knew about germs and had understood fermentation before Pasteur. In the euphoria of Pasteur's glory Béchamp dared to say: 'Instead of trying to determine what abnormal conditions disease is composed of, let us first know the normal conditions which make us healthy'. His understanding of health and diseases can easily be converted into a twenty-first-century scientific way of thinking.[24]

CHAPTER SIX

Should we criminalize planned vaginal birth?

We have offered an overview of scientific advances that will undoubtedly influence the history of childbirth, at a time when the behavioural effects of hormones involved in childbirth are widely documented and which is also the age of epigenetics and the microbiome revolution. This overview has inspired futuristic questions that may be considered marginal, as it is still unusual to merge the history of childbirth and the history of *Homo sapiens*.

We cannot escape overnight from the world in which we live and from our deep-rooted cultural conditioning. One can easily understand why, for thousands of years, a birth has been considered successful if mother and baby are alive and healthy. Today, medical language still expresses the same way of thinking. The usual criteria for evaluating the practices of obstetrics and midwifery are the perinatal mortality rates (the number of babies who die before the age of a week), the perinatal morbidity rates (in practice, the number of newborn babies who need to be transferred to a paediatric unit), the maternal morbidity rates (maternal health problems related to childbirth) and maternal mortality rates. Despite recent spectacular technical advances, this short list of criteria has not been modified. We must keep in mind that the caesarean section has become an easy and fast operation, usually performed with regional anaesthesia. It is worth recalling the two important steps in the history of caesarean techniques.

Two important steps

When I did my first caesarean sections in the 1950s, general anaesthesia (with the methods of that time) was routine. We usually needed an hour to finish the operation. We also needed one or two bottles of blood to compensate for the blood loss. Women stayed in hospital for a week after what was considered a major abdominal operation. At that time we had already taken the first important step in the history of caesarean techniques. We were opening the uterine muscle by a transverse (side to side) incision at a thin zone called the low segment. The risks of all sorts of complications were already dramatically reduced. Before then the most direct route was used to open the uterus. The body of the uterus was cut by a vertical (up and down) incision. For many reasons this 'classical' section was done as a last resort. The risks of bleeding from the thick uterine wall and the risks of infection were too high; bowel adhesions to the uterine scar could cause intestinal obstruction; healing of the uterine muscle was often faulty and the risk of bloody scar rupture in subsequent pregnancies was high.

The second important step came in the 1990s, when Michael Stark and the team at the Misgav Ladach Hospital, in Jerusalem, introduced methods that restricted the use of sharp instruments, preferring manual manipulation instead. One of the objectives is to remove every unnecessary step. The advantages, in terms of speed and blood loss, are spectacular. Today, the average operating time is in the region of twenty minutes and the average blood loss is roughly the same as after a vaginal birth. Mothers and babies go home within forty-eight hours.

Measuring the safety of caesareans

In general, those who are directly involved in obstetrics regard the modern caesarean as a safe operation. It is significant that many women obstetricians choose to have a caesarean for the birth of their own babies.[1,2] It is also noticeable that all over the world, more and more women simply choose to give birth by c-section, particularly in emerging countries such as China and Latin America. Of course, it is difficult to express the degree of safety in statistical language. The gold standard method of medical research to compare two possible treatments, or policies, or strategies, is based on randomization: this means that a population is first divided into groups by drawing lots. One group is allocated a treatment, while another group is allocated – at random – another treatment. Then, during a follow-up period, the comparative ratios of benefits to risks for both treatments are evaluated and expressed in statistical language. For obvious reasons one cannot tell a group of pregnant women that they must give birth vaginally, while other women – at random – are told that they must have a caesarean.

However, we can learn from retrospective studies. A typical example is a large Danish study[3] looking in retrospect at the 15,441 Danish women who gave birth to a first baby in a breech position between 1982 and 1995. Among them 7,503 had a planned caesarean, 5,575 had an emergency caesarean, and 2,363 gave birth vaginally. Contrary to received ideas, the incidence of haemorrhage and anaemia after planned caesarean section (5.7%) did not differ from that after vaginal delivery and was slightly lower than after emergency caesarean (7%). The rate of thromboembolic complications was 0.1% after caesarean. On the other hand the rate of anal sphincter rupture, which is associated with a subsequent risk of anal incontinence, was 1.7% after vaginal birth. It is commonplace to emphasise that all surgical procedures carry an inherent risk of injuries to organs not directly involved in the particular surgery. The risks must be put in perspective. In this study there were just a few cases of bladder injuries (0.1% during planned caesarean and 0.2% during emergency caesarean), and bladder injuries can be easily and immediately repaired. The risks of such injury are very low with techniques that restrict the use of sharp instruments. In this series of births it was never necessary to perform a 'caesarean-hysterectomy'(to remove all or part of the uterus) to stop the bleeding. The risks of post-operative adhesions (and therefore the risks of intestinal obstruction long after) are also very low after the modern caesarean, particularly if no foreign material such as gauze has been introduced to the abdomen.

The risks of maternal death are particularly difficult to evaluate. Once more the studies cannot be randomized. In most statistics these risks appear three to four times higher after a caesarean than after a vaginal birth.[4] But the studies are hampered by the fact that the women having caesarean delivery have conditions, pregnancy complications, and delivery complications that are themselves associated with increased maternal mortality. Furthermore, in developed countries, one needs to analyse the outcomes of at least 100,000 births to significantly evaluate the rates of maternal deaths (or 'pregnancy-related deaths'). These difficulties are to a certain extent eliminated now since, in many hospitals in industrialized countries, the routine is to perform planned c-sections at thirty-nine weeks in the case of a breech presentation at term. Such strategies enable comparison of huge numbers of c-sections that are not related to maternal pathological conditions. In a report concerning all women in Canada (excluding Quebec and Manitoba) who gave birth between April 1991 and March 2005, the planned caesarean group for breech presentation comprised 46,766 women.[5] No mothers died in this group, while forty-one mothers died in the 2,292,420 women in the planned vaginal delivery group (maternal mortality rate 1.8 per 100,000 births: the differences are not statistically significant).

Of course there are now documented cases of abnormal pregnancies after caesareans (in particular increased risks of abnormal placenta insertions, and caesarean scar ectopic pregnancies). Such risks must be put into perspective, at a time when, in many developed and emerging countries, the average number of babies per woman is below two.

In general, whatever the perspective, it is indisputable that in modern well-equipped and well-organized hospitals the caesarean has reached a degree of safety that is impressive to practitioners of my generation. We might even wonder why routine c-sections at thirty-nine weeks are not officially promoted. A study considered that among the four million births each year in the USA, about three million occur at thirty-nine weeks or after. It was estimated that if all American women reaching thirty-nine weeks gave birth by caesarean, either before labour started or once labour was established, 9,462 cases of neonatal encephalopathy a year would be prevented.[6] In reference to the conventional criteria we mentioned – particularly perinatal mortality and morbidity rates – it is currently commonplace to denigrate out-of-hospital births. There is even a tendency today in some countries towards a quasi-criminalization of home births. As long as the list of criteria to evaluate the practices of midwifery/obstetrics remains the same, it would be more rational to ban planned vaginal birth. Why not?

CHAPTER SEVEN

That is the question

There have been recently countless published documents and discussions comparing the 'safety' of hospital births and the 'safety' of out-of-hospital births. These discussions have involved a great diversity of milieus, from the most conservative specialized medical circles to the promoters of 'natural childbirth'. The point is that all these groups share the same battlefield. In other words they all refer to perinatal mortality and morbidity rates and to maternal morbidity and mortality rates. They limit their discussions to the place of birth. Why does nobody, in the light of the evidence, dare to encourage routine caesarean birth? That is the question.

We just need to phrase the question that way to realize that the simplified techniques of caesareans have created an unprecedented situation. It is as if a widespread intuitive piece of knowledge is imposing limits on the dominant way of thinking at a phase of history when it should be considered rational to advise all pregnant women to give birth by caesarean.

Answer

The answer is clear: the time has come to enlarge the list of criteria evaluating how babies are born. The widespread intuitive knowledge that is slowing down the tendency towards generalized caesarean births is not strong enough to go a step further. The support of scientific evidence is vital. This is why we have presented an overview of scientific advances that might influence the history of childbirth. This is the case in particular of the studies demonstrating the behavioural effects of hormones involved in childbirth, and also of the concepts of gene expression and early bacterial colonization of the newborn's body.

While the scientific disciplines we have mentioned offer plausible interpretations and hypothetical long-term consequences of what happens at birth, we must wonder if there are already published hard data demonstrating such consequences.

The answer is provided by an exploration of the Primal Health Research Database (www.primalhealthresearch. com). Primal health research may be presented as a specialized branch of epidemiology. It explores correlations between what happens during the primal period (foetal life, perinatal period, and the year following birth) and what happens later on in terms of health and personality traits. Of course the epidemiological perspective can only detect correlations that are presented in statistical language. However, when the findings are statistically significant, when the results of several studies are concordant, and when there are plausible interpretations, the cause and effect relationship is semi-certain. This is why, when considering the accumulation of available data, we can claim that the long-term consequences of how we are born have been underestimated until now. The constant interaction of primal health research and emerging scientific disciplines that suggest interpretations of epidemiological results explains the vitality of our database.

A reminder of the history of the concept of primal health research is necessary to understand the multiple functions of the Primal Health Research Database.

In July 1982 I was invited to speak in Oxford at a conference organised by the McCarrison Society. My presentation, entitled 'Childbirth and the diseases of civilization', was a plea for a new kind of research to test the hypothesis that our health is, to a great extent, shaped at the very beginning of our life. We already had theoretical reasons to suppose that the sudden increased incidence of certain pathological conditions might be related to new, powerful ways of interfering with the physiological processes in the perinatal period. At this conference I met Niko Tinbergen, a founder of ethology, who had won a Nobel Prize in 1973. He was the pioneer I was looking for, since he had explored the risks of autism in relation to how the child was born. The correspondence we had afterwards encouraged me to write a book about the concept of health and what makes human beings more or less healthy.

This is how, in 1986, I published the first edition of *Primal Health*.[1] Back then we could already predict that the spectacular advances in computer sciences would facilitate the new generation of research we were waiting for. It would suddenly become easier to explore correlations between what happens at the beginning of life and what happens later on.

I was convinced that we had to prepare for this new generation of research, first by adapting our vocabulary. I had to give a definition of 'the beginning of life'. The period of human development when the basic adaptive systems – those involved in what we commonly call health – reach their maturity was called the 'primal period': it includes foetal life, the perinatal period, and the year following birth. I also underlined the need for a simple term in order to get rid of the artificial and obsolete separations between the nervous system, the immune system and the endocrine system. I suggested the term 'primal adaptive system' when referring to this network: a way to avoid awkward terms such as 'psycho-neuro-immuno-endocrinologic system'. 'Health' is how well our primal adaptive system works. 'Primal health' is the basic state of health in which we are at the end of the primal period; after that we can take advantage of and cultivate this basic state of health. 'Primal health research' is this new framework of studies exploring correlations between what happens during the primal period and what will happen later on in life.

After the publication of *Primal Health*, the next step was to create, in 1987, the Primal Health Research Centre, based in London. Apart from occasional studies we could do by ourselves, our objective was to collect medical and scientific literature within the framework of primal health research. An explosion of such studies came in the late 1980s. At the beginning I was just collecting and classifying printed articles.

It was in around 1990 that variants of the hypotheses included in *Primal Health* appeared in the mainstream medical literature. The 'foetal origins of disease hypothesis' was inspired by the countless studies published by the Barker group, a team of British epidemiologists based in Southampton. It became the 'foetal/infant origins of disease hypothesis' in the mid-1990s.[2] There are obvious similarities between the 'studies testing the foetal/infant origins of disease hypothesis' and primal health research. One of the main differences is that our key word is 'health' instead of 'disease'. Improving our understanding of health may prove more fruitful than studying the origins of particular diseases.

In 1993 I was asked to prepare a second edition of *Primal Health*. I thought it was impossible to update a book about such a fast-evolving field of research. It was more relevant to launch a quarterly newsletter to provide updated information to a small core of supporters who had already understood the importance of this new generation of research. Each issue takes the form of an essay on one particular subject relevant to primal health research. Some of these essays have been modified in order to be included on www.wombecology. com: the objective of this website is to convince anyone that pre- and perinatal ecology is the most vital aspect of human ecology, and that the period inside the womb is the life period with the greatest adaptability and vulnerability to environmental factors.

In the late 1990s the spectacular advances in information technology and the fast development of primal health research led us to create a database available on the web. Today we need this database because it is difficult to identify – among thousands of scientific and medical journals – studies within this new framework: they are unrelated according to the current classifications. For example, it would take a long time to find out that a pregnancy disease such as pre-eclampsia has been studied in relation to health conditions as diverse as prostate cancer, breast cancer, hypertension, asthma, allergic rhinitis, type-1 diabetes, body size, age at menarche, behaviour disorder, schizophrenia, mental retardation and cerebral palsy. After the Primal Health Research Database was established, I agreed to republish the book *Primal Health* without any alteration to the original text. We just added an introduction to the second edition.[3] This new edition was a useful document for evaluating the progress in primal health research over two decades. The original 1986 chapter 'Research in Primal Health' was followed by only fifteen references!

Today the database contains hundreds of studies that have been published in authoritative medical and scientific journals. We offer hundreds of keywords in order to facilitate access to the relevant references and abstracts. In all fields of medicine and health sciences there are already studies that belong to the framework of primal health research.

Multiple functions of the Primal Health Research Database

The primary function of our database is to train ourselves to think long-term. Human beings have not been programmed for long-term thinking. This is not special to midwives and obstetricians. For millions of years our tropical ancestors consumed the food they could find from day to day in their environment, either by collecting shellfish and small fish in shallow water, by gathering plants and fruits, or by scavenging and hunting. After the comparatively recent advent of agriculture and animal breeding, they had to increase their capacity to anticipate. They were obliged to think at least in terms of seasons. Today we have at our disposal such powerful technologies that we must suddenly condition ourselves to think in terms of decades, centuries and millennia. This is the only way to prevent major conflicts between humanity and Mother Earth. We must start by raising questions about childbirth.

To think long-term is a way to enlarge our horizon when evaluating different ways of being born. There are also reasons to enlarge our horizon by thinking in terms of civilization, a specifically human dimension: our database can also be presented as a tool for becoming familiar with this collective dimension, since epidemiologists often need huge numbers to detect statistically significant effects of early experiences. A typical example, among many others, is provided by one of the studies of the risk factors for autism. The researchers had at their disposal the birth records of the entire Swedish population born during a period of twenty years; they also had the medical files of all Swedish subjects who had been diagnosed as autistic with strict criteria, plus five controls for each of them. With such material they could detect statistically significant risk factors in the perinatal period. (See entry 0396 via keyword 'autism'.)

The need to introduce the collective dimension and to think in terms of civilization appears clearly when considering the effects of disturbing the birth process of non-human mammals. Among other mammals, when the birth process has been disturbed, the effects are spectacular and easily detected immediately at an individual level: the mother is not interested in her baby. This is the case with ewes giving birth with epidural anaesthesia[4] or monkeys giving birth by caesarean.[5] Their young can survive only if human beings take care of them. Millions of women, on the other hand, take care of their newborn babies in spite of powerful interferences.

We can easily understand why it is much more complex in our species. Because human beings speak and create cultural milieus, there are situations when our behaviour is less directly under the effects of the hormonal balance and more directly under the effects of the cultural milieu. This is the case in pregnancy and childbirth. When a woman is pregnant, she can express through language that she is expecting a baby and she can anticipate maternal behaviour. Other mammals cannot do that. They have to wait until the day when they release a cocktail of love hormones to be interested in their babies. We should not conclude that we have nothing to learn from other mammals. They suggest which questions we should raise where human beings are concerned: in these questions we must always introduce the collective dimension via words such as 'civilization'. This is why today the main questions are about the future of civilizations born by caesarean, or with epidural anaesthesia, or with drips of synthetic oxytocin…

The main lesson of the primal health research perspective is that we should interpret anecdotes with extreme caution. With anecdotes one can support any thesis. Thinking in terms of civilization also means that those who explore epidemiological studies must first forget their family, their friends, and particular cases as well. We should not be worried about one particular baby who was rescued by caesarean. The cultural milieu can to a great extent compensate many deprivations. The questions must be raised differently when human beings are concerned.

The advent of epigenetics suddenly gave the Primal Health Research Database another function. From our database we understand the importance of the time of exposure to an environmental factor. For example, when exploring our database from a keyword related to a metabolic type (such as 'type-2 diabetes' or 'obesity'), foetal life usually appears as a critical period for gene-environment interaction. If the keyword is related to the capacity to love it appears that the period surrounding birth is critical. Our database has become a unique tool to provide some clues about the critical periods for the genesis of diseases and personality traits. From a practical perspective it is often more important to identify such critical periods than to identify the genes involved or the comparative roles of genetic and environmental factors.

We might illustrate this function of our database by taking asthma as an example. The well-documented increased incidence of asthma is still partly mysterious. It occurred at such a speed, within just a few decades, that it cannot be explained by genetic factors. Recently, a study of DNA methylation (as an epigenetic marker) at the level of one asthma candidate gene (IL4R) has added to the explanation for asthma. One can conclude from the study that epigenetic processes modulate the risks of asthma.[6] This piece of knowledge inspires questions in terms of timing. The first question is about the critical period for gene-environment interaction. An exploration of the Primal Health Research Database indicates the importance of the period surrounding birth. A caesarean birth and exposure to antibiotics, in particular, appear as risk factors. These facts open the way to interpretations of the nature of the link. Is asthma, as a respiratory disease, related to the frequent respiratory difficulties experienced by newborn babies after a pre-labour caesarean? Is asthma, as an allergic disease, related to an altered bacterial colonization of the digestive tract, which is the rule after a birth via the abdominal route and also after exposure to antibiotics? Our database has a central role to play in solving the puzzle.

The paramount importance of the concept of 'timing' explains my personal interest in pathological conditions that share the same critical period for gene-environment interaction. It appears that when two diseases share the same critical period, other similarities are probable, such as when viewed from clinical and physiopathological perspectives. This is how I have explored, for example, the multiple links between autism and anorexia nervosa.[7] Can the Primal Health Research Database inspire new ways to classify diseases?

The concept of critical period

In the age of epigenetics, the Primal Health Research Database is giving renewed importance to the concept of critical (or sensitive) periods in the development of living creatures and particularly human beings. This concept was originally introduced at the beginning of the twentieth century by the Dutch geneticist Hugo De Vries. It was adopted by Maria Montessori to refer to short yet important periods of childhood development. Maria Montessori was referring mostly to the first years of life, from birth (and before) to the time of what she considered complete development of the brain, about age six or seven.

Since that time the concept of critical period has been widely used by a great diversity of scientists, particularly ethologists. The ethological critical period concept is based on the fact that an individual's behaviour is more strongly influenced by an event at one stage of development than at others. For example there is a critical period during which birds imprint upon their mother and learn their species' songs. The concept was popularized in the 1930s with an experiment which has become a legend. The founder of modern ethology, Konrad Lorenz, reported that one day he had interposed himself between newly-hatched ducklings and their mother and then imitated the mother duck's quacking sounds. These ducklings became attached to Lorenz for the rest of their lives, following him when he walked in the garden, for example. This is how the concept of a critical (or sensitive period) in the process of forming attachment was introduced, and it shows that there is a short yet crucial period immediately after birth which will never be repeated. The ethological perspective inspired human studies evaluating the effects of immediate skin to skin contact between mother and newborn baby. It also inspired studies of the behavioural effects of hormones that fluctuate in the period surrounding birth. Such studies suggested interpretations of the concept of critical period in mother-baby attachment. It became easy to explain that, among humans, it takes about an hour for maternal and foetal hormones to be eliminated after the birth process and that each hormone has a specific role to play in the interaction between mother and baby.

Today, the microbiome revolution is offering other reasons for revisiting the concept of critical period, since being born is entering the world of microbes, and since great importance is given to the initial phase of the microbial colonization of the newborn's body.

At the same time, in the age of epigenetics and primal health research, the concepts of gene expression and of gene-environment interaction are inspiring a new generation of studies of the trans-generational effects of early experiences.

In such a scientific context there is no doubt that the period surrounding birth is crucial in the development of human beings. It is also a phase of human life that has been radically transformed by modern lifestyles. All those interested in the future of *Homo sapiens* should be interested in the way babies are born.

CHAPTER EIGHT

Active management of human evolution

Can we and should we consciously direct the transformation of *Homo sapiens*?

The reasons for a new question

In the current scientific context, at a time when the survival of our species is at stake, and at a time when there are multiple ways of being born, we must dare to raise this question.

This would be the most important step in the domination of nature by the genus *Homo* since the Neolithic revolution, when our ancestors started to domesticate plants and animals and to practice agriculture and animal husbandry.

Of course, since the Neolithic revolution, cultural milieus have interfered with the birth process and have undoubtedly transformed *Homo sapiens*. Human groups we know about have transmitted countless perinatal beliefs and rituals from generation to generation, the effects of which are to amplify the difficulties of childbirth, to separate mothers and newborn babies, and to delay the initiation of breastfeeding. But the objective has never been to consciously transform *Homo sapiens*. The transformations of *Homo sapiens* that followed the Neolithic revolution were the effects of a kind of natural selection between human groups.

We must realize that from the time when the basic strategy for survival of most human groups was to dominate Nature and to dominate other human groups, it was an advantage to develop the human potential for aggression and the capacity to destroy life. In other words it was an advantage to moderate the capacity to love, including love of Nature, or respect for Mother Earth. If the perinatal period is critical for gene-environment interaction regarding the development of the capacity to love, and inversely the potential for aggression, it was an advantage to interfere in the birth process, particularly the third stage of labour and the first contact between mother and baby. Over time there has been a selection of human groups according to their capacity to develop the potential for aggression.[1] '*Homo superpredator*', the current variant of *Homo sapiens*, is the fruit of such a selection. *Homo superpredator* appeared when our ancestors turned the strategies for survival upside-down. They started to transform the environment and to adapt it to their needs. Up to that time, all animals, including human primates, had survived by adapting to an environment.

How can respect for Mother Earth – a facet of love – develop? How can we unify the planetary village? Raising such new questions would be a way to introduce the concept of active management of human evolution. The first step would probably be to challenge those who believe that there are many other factors to consider than the way babies are born. This is why it is necessary to focus on the concept of critical periods in the development of human beings and to realize that the period surrounding birth is the episode of human life that has been the most completely altered for thousands of years, particularly during recent decades. The second step would be to agree on objectives.

Active management needs objectives

Many advocates of the futurist and transhumanist movements dream of transcending the limits of human potential. They dream in particular of an evolution towards a higher degree of encephalization. An increased development of the neocortex should be associated with an improved average intellectual potential, and therefore an increased capacity to solve difficult problems related to the future of *Homo sapiens* and the future of the planet. It seems that these theoreticians, who are versed in a great variety of emerging scientific disciplines, don't realize that the caesarean has become an easy and fast operation. They don't realize that the widespread availability of caesarean births implies that the tendency towards an increased head circumference can be transmitted to the following generations without any restriction. Today evolution towards a higher development of the neocortex is not utopian. We assume that many futurists will welcome these possible consequences of the simplified techniques of caesarean.

Others might fear the drawbacks of an excess of rationality if the evolution towards a more powerful neocortex is accompanied by a weakening of basic physiological functions that are under the control of archaic brain structures. We have already expressed this concern in relation to the oxytocin system, which is involved in all facets of the capacity to love and is essential for the survival of the species. The point is that the will to survive and the struggle for life are not rational. The desire to have descendants and the need to take care of them is not rational. The many facets of love are outside the field of rationality. Can a super-brainy *Homo sapiens* be endowed with a strong 'emotional intelligence'? Can a super-brainy *Homo sapiens* survive without love?

There is food for thought in the accumulation of data supporting de Catanzaro's evolutionary theory of human suicide,[2] according to which a threshold intelligence is necessary for self-damaging behaviour, which might be presented as an impaired capacity to love oneself. The huge statistical studies by Martin Voracek in eighty-five countries confirm that a high intelligence quotient (IQ) is a risk factor for committing suicide.[3] Similar conclusions are provided by the famous Terman genetic study of genius. This still-running longitudinal study of 1,528 gifted Californian children born in 1920–25 revealed that among super-bright individuals the rate of suicide is three times higher than for the rest of the American population.

In general, one might consider it risky to routinely and consciously manipulate the comparative development of archaic and recent brain structures. This is the lesson provided by emerging disciplines confirming the function of stress in the perinatal period on brain development, particularly the development of the hippocampus.[4] Let us recall the importance of the hippocampus in memory and its deterioration in Alzheimer's disease.

Should we conclude that caesarean birth must be avoided at any price? This is a common point of view among some natural childbirth groups: they consider 'caesarean prevention' a priority. This point of view is indirectly supported by recommendations frequently expressed by public health organizations.

This would be another risky attitude. In a mainstream medical journal I have expressed a warning about the negative effects of reducing the rates of caesarean section as a primary objective.[5] The most dangerous guidelines are those suggesting a limit on the number of caesarean sections. The first expected effect is to increase the rates of difficult instrumental vaginal deliveries with pharmacological assistance, which should become exceptionally rare in the age of the safe caesarean. This effect is already detectable in countries such as the UK, where the rate of forceps has doubled during the first decade of the twenty-first century. Common sense, supported by studies included in the Primal Health Research Database, suggests that an in-labour-non-emergency caesarean section should often be preferred to a birth after hours of drips of synthetic oxytocin and epidural anesthesia.[6] Most babies born by in-labour-non-emergency caesarean are particularly alert, compared with those born by caesarean before the labour starts and those born by last-minute emergency caesarean. While the negative side-effects of forceps and ventouse deliveries, and also of epidural anaesthesia during labour, are well documented, it seems that the side-effects of synthetic oxytocin have been underestimated and even ignored until now. According to preliminary studies, the drips of synthetic oxytocin during labour (by far the most common medical intervention in childbirth) interfere in particular with lactation and also with the psychomotor development of the child.[7] There are several possible (and complementary) interpretations of the negative effects of synthetic oxytocin on the initiation, the quality and the duration of breastfeeding. One of them is a desensitisation of the oxytocin breast receptors. Another one is an effect on the maternal hypothalamo-pituitary axis. Another one is that a drip of synthetic oxytocin is usually associated with an epidural anaesthesia, which independently has a negative effect on breastfeeding. As we have seen, there are serious reasons to claim that the high concentrations of synthetic oxytocin in maternal blood can cross the placenta, reach the foetal brain across an immature blood-brain barrier and eventually alter the behaviour of the newborn baby.[8] This seems to be confirmed by video studies of the primitive reflexes of the newborn baby, which are depressed after this type of pharmacological assistance.[9] It is also confirmed by videotapes of infants forty-five to fifty minutes after birth: fewer breastfeeding cues were observed among infants exposed to synthetic oxytocin.[10]

Finally, we suggest that, in the framework of an active management of human evolution, the objective should be expressed in a positive way. It should be our aim to create the right situation for as many women as possible on this planet to give birth to babies and placentas thanks to a flow of love hormones. Such an objective implies that the basic needs of labouring women are rediscovered. Is it utopian?

CHAPTER NINE

Physiology v cultural conditioning

Understanding the laws of Nature

Since the Neolithic cataclysm – around 10,000 years ago – the basic strategies for survival of human groups have been to dominate Nature. This domination of Nature went beyond the domestication of plants and animals. It has included also the control, the organization, and often the repression of human physiological processes, particularly those related to reproduction. There are numerous ways today to make clear that we are fast approaching the limits in all aspects of the domination of Nature. It is common and accepted knowledge where climate change and air and water pollution are concerned. It is not yet well understood where childbirth is concerned, although in this case it is more than a control of the physiological processes: it is often a real replacement by pharmacological and/or surgical methods. In the case of childbirth, it is the evolution of *Homo sapiens* that is directly at stake. The issues related to the future of the planet are secondary, since the transformations of the planet are first and foremost dependent on the evolution of *Homo sapiens*.[1]

In all aspects of human activity, there is an urgent need to improve our understanding of the laws of Nature in order to work with them. Where childbirth is concerned, this means that our point of departure should be the physiological perspective. Thanks to this perspective it is possible to analyze basic and universal human needs and to transcend cultural particularities.

It is easy to summarize how the birth process is understood in the framework of modern physiology. The birth process appears as an involuntary process under the control of archaic brain structures (such as the hypothalamus and pituitary gland) that are the sources of an appropriate hormonal flow. This fact inspires the comment that, in general, one cannot help an involuntary process, but there are situations that can inhibit it. Thanks to established concepts, such as the antagonism between adrenaline and oxytocin and neocortical inhibition, modern physiology has the power to identify such situations. In other words, from this perspective, the keyword is 'protection' (of an involuntary process).

Deep-rooted cultural conditioning

There is a striking contrast between this vision of the birth process offered by modern physiology and our deep-rooted cultural conditioning. For thousands of years, the basis of our cultural conditioning regarding childbirth has been that a woman is unable to give birth by herself. She has not the power to give birth without some kind of cultural interference.

An analysis of language is an effective way to explore our cultural conditioning. 'Who delivered your baby?' is a common question asked of mothers. The expected answer is not 'I gave birth'. The international modern word 'obstetrics' and the Latin word for 'midwife' ('obstetrix') come from 'ob-stare': it implies somebody must be positioned in front of the labouring woman. This disempowering vocabulary is not confined to Western languages. In Chinese they often use the term '*jie Sheng*', which literally means 'delivery carried out by others'. Recently introduced terms have not toned down the effects of older words. Mothers-to-be are 'patients': it is commonplace to contrast the patients (passive) and the care-providers (active). Today we often hear about 'methods' of 'natural' childbirth, as if the words 'method' and 'natural' were compatible. The word 'method' implies that the labouring woman must follow a pattern pre-established by an expert.

An analysis of rituals transmitted from generation to generation is another way to investigate cultural conditioning. We just need to consider some of the most widespread rituals to notice that one of their effects is to make absolutely necessary the active participation of an agent of the cultural milieu. This is the case with ritual genital mutilations, which were reported by Herodotus in ancient Egypt in the fifth century BC. The practice of female genital mutilation is found across Africa in a broad triangular east-west band that stretches from Egypt and Tanzania in the East up to Senegal in the West. Most of these operations (excisions and infibulations) leave a rigid scar: somebody must be there when a woman is in labour in order to cut the perineum and make birth possible. Early cord-cutting has been another widespread ritual, for thousands of years. It is associated with the belief that at the very time when the baby is born, somebody must be there to do something. This belief is still strong in our societies. There are anecdotes about men who were in a panic while their wife was giving birth before the midwife arrived, because they did not know how to cut the umbilical cord.

Reinforced cultural conditioning

Until recently beliefs and rituals were the main factors influencing cultural conditioning. From the middle of the twentieth century theories started to play an important role. Among the most influential theories we must mention those of Pavlov, associated with the concept of conditioned reflexes. These theories led to the advent of

'Psychoprophylactic Methods' and many schools of 'natural childbirth'. The final effect has been that women must learn to give birth and need to be continuously guided during labour and told how to breathe, how to push, and so on. This is how an unprecedented and sophisticated form of culturally controlled childbirth suddenly developed. This is how the cultural conditioning has been dramatically reinforced.

Today we are mostly conditioned by visual messages. We are in the age of videos, photos, television and the internet. There is a real epidemic of videos and photos of childbirth. On these pictures, there are almost always at least two or three persons surrounding the labouring woman. The message perceived by younger generations is clear. It is that the basic need of a labouring woman is to be accompanied by other people who bring their expertise or their energy. The visual messages are reinforced by the modern vocabulary. A 'coach' is bringing her expertise. A 'support person' is bringing her (his) energy. The basic message is, more than ever: 'you cannot give birth without the energy or the expertise of others'. We have reached another phase in the history of our cultural conditioning.

We must add that this cultural conditioning is now shared by the world of women and the world of men as well. While traditionally childbirth was 'women's business', men are now almost always present at births, at a phase of history when most women cannot give birth to the baby and to the placenta without medical assistance. A whole generation of men is learning that a woman is not able to give birth. We have reached an extreme degree in terms of conditioning.

The current dominant paradigm has its keywords: helping, guiding, controlling, accompanying, managing ('labour management'), coaching, supporting… the focus is always on the role of other persons than the two obligatory actors (i.e. mother and baby). Inside this paradigm, we can include women, men, medical circles and natural childbirth movements as well. It is cultural.

Will twenty-first-century scientific disciplines be powerful enough to induce a new awareness and a vital paradigm shift?

CHAPTER TEN

Reasonable optimism

Concern about the future of *Homo sapiens* is reasonable. This is our conclusion after considering the limits recently reached in the history of childbirth. However there are also reasons for optimism. The main reason for optimism is the capacity of modern scientific perspectives to challenge thousands of years of cultural conditioning and to stimulate common sense via simple conclusions.

Before a spectacular scientific discovery

This power can be illustrated by a spectacular discovery of the second half of the twentieth century. Until that time nobody knew that a newborn baby needs its mother. When I was an 'externe' (medical student with minor clinical responsibilities) in a Paris hospital, in 1953, I never heard of a mother who would have said, just after giving birth: 'Can I keep my baby with me?'. The cultural conditioning was too strong. Everybody was convinced that the newborn baby urgently needed 'care' given by a person other than the mother. The midwife was quick to separate mother and baby by cutting the umbilical cord and would put the baby in the hands of a nurse. This is what she had learnt to do at the midwifery school. At that time it would have been the same in the case of a home birth. Then, while staying in the maternity unit, babies were in nurseries and mothers were elsewhere. Mothers were not asking to stay in the same room as their baby.

We must realize that for thousands of years, in all human societies we know about, mothers and newborn babies have been separated and the initiation of breastfeeding has been delayed. In other words it has been routine for a long time to neutralize the 'maternal protective aggressive instinct'. The nature of this universal mammalian instinct is easily understood when one imagines, for example, what would happen if one tried to pick up the newborn baby of a mother gorilla who had just given birth.

It would take volumes to review all the invasive perinatal beliefs and rituals that have been reported in a great variety of cultures. As early as 1884 *Labor Among Primitive Peoples* by George Engelmann provided an impressive catalogue of the one thousand and one ways of interfering with the first contact between mother and newborn baby. It described beliefs and rituals occurring in hundreds of ethnic groups on all five continents.[1]

The most universal and intriguing example of cultural interference is simply to promote the belief that colostrum is tainted or harmful to the baby, and that it is even a substance which needs to be expressed and discarded.[2] The negative attitude towards colostrum implies that, immediately after the birth, a baby must be in the arms of another person, rather than with his or her own mother. This is related to the widespread deep-rooted ritual of rushing to cut the cord.[3] Several beliefs and rituals can be seen as part of the same interference, all of them reinforcing each other.

Western Europe is not a stranger to these universal rules. In Tudor and Stuart England, colostrum was openly regarded as a harmful substance to be discarded.[4] The mother was not considered 'clean' after childbirth until the bloody discharge called lochia had stopped flowing. She was not permitted to give the breast until after a religious service of purification and thanksgiving called 'churching'. Meanwhile the baby was given a purgative made from such things as butter, honey and sugar, oil of sweet almonds or sugared wine. Paintings from that time show the newborn infant fed with a spoon while the mother recovered in bed. In Brittany the baby was not put to the breast before baptism, which took place at the age of two or three days. The Bretons of old believed that if the baby swallowed milk before the ceremony, the devil might enter the baby's body along with the milk.

The discovery

Recalling these roots of our cultural conditioning is a necessary step in evaluating the importance of the scientific advances of the 1970s. A new generation of human studies was inspired by what we learned about mammals in general. It is significant that Konrad Lorenz, Nikolaas Tinbergen and Karl Von Frisch were the joint winners of the Nobel Prize in Physiology and Medicine in 1973. It is thanks to these founders of ethology, and the work of other ethologists such as Klopfer,[5] that we became familiar with the concept of critical periods for mother-newborn attachment. In other words we understood that among mammals in general, immediately after birth, there is a crucial short period of time that will never happen again. This was the beginning of what we have called 'The Scientification of Love'.

The time was ripe to evaluate the effects of immediate body-to-body contact between mother and newborn baby as an absolutely new intervention among humans. The names of Marshall Klaus and John Kennell in the USA are associated with such studies,[6] which were also conducted in Sweden.[7,8,9] In parallel other researchers were interested in the behavioural effects of hormones that fluctuate in the perinatal period, particularly oestrogens.[10,11,12] This was also the decade in which a sudden interest in the content of human colostrum developed. Until that time 'colostrum' was a fruitful keyword in veterinary medicine, but not in human medicine. In the seventies the focus was on local antibodies (IgA) and anti-infectious substances.[13,14,15] After thousands of years of negative connotations human colostrum was officially recognized as a precious substance.

In the 1970s we also learned that when there is a free, undisturbed and unguided interaction between mother and newborn baby during the hour following birth, there is a high probability that the baby will find the breast during that time: human babies usually express the 'rooting reflex' (searching for the nipple) during the hour following birth, at a time when the mother is still in a special hormonal balance and has the capacity to behave in an instinctive 'mammalian' way. The result of the complementary behaviour between mother and newborn baby is an early initiation of breastfeeding.[16,17] For obvious reasons, nobody knew, before the 1970s, that the human baby was programmed to find the breast during the hour following birth.

The 1970s was also a period of rapid development in immunology and bacteriology. We have mentioned the importance of studies about the easy and effective transfer of maternal antibodies (IgG) across the human placenta.[18,19] This implies that the microbes familiar for the mother are also familiar, and therefore friendly, for the germ-free newborn baby. We had reached a new vision of human birth from a bacteriological perspective. We were in a position to understand that the main questions are about the first germs that occupy the territory and become the rulers of the territory. In other words, we were in a position to understand that from immunological and bacteriological perspectives a newborn baby needs to be in urgent contact with the only person with whom he (she) shares the same antibodies (IgG).

After referring to these extensive scientific activities of the 1970s, we can observe that it has been possible, during the second half of the twentieth century, to discover the basic needs of the newborn baby. We summarize these basic needs by claiming that the newborn baby needs its mother.

Thanks to sophisticated modern scientific perspectives, it is becoming difficult to ignore the fact that we are mammals.

Immediate implications of the discovery

The discovery that the newborn baby needs its mother immediately had practical implications. It is not by chance that the concept of 'rooming in' suddenly developed: it implies that mother and baby are in the same room while staying in the maternity unit. In fact this concept originally appeared in some American hospitals during the Second World War. However, at that time, the main reasons for this new organizational arrangement were practical. It was a way to adapt to the nursing shortage. In some hospitals it had been a way to reduce the effects of epidemics of neonatal infections in nurseries. This concept spread out all over the world. In 1981, I was invited to speak at a conference about 'rooming in' in Olomouc, Czechoslovakia. At that time, in the context of the Communist regime, being critical of the concentrations of neonates in nurseries was considered revolutionary. It was also in 1981, in the same scientific context, that I went to Bogotà, Colombia, where they had already been practicing 'kangaroo care' for premature babies for two years (although the term kangaroo care was not used at that time). They had understood that when a premature baby does not need assistance to breathe the mother can be the best possible incubator.

The effects of these scientific advances went beyond the period surrounding birth. It was also in the late 1970s that baby carriers became suddenly fashionable. This was also the time when breastfeeding was promoted again. Breastfeeding had been devalued in the middle of the twentieth century due to various factors, such as the development of the food industry and the advent of 'humanized milk', and also theories expressed by some feminist movements.

Of course, one can easily understand why it has been difficult to reverse overnight the effects of thousands of years of cultural conditioning. In spite of scientific advances it has been difficult to accept a free interaction between mother and baby, without any cultural interference, immediately after birth. It is significant that, while scientists were explaining that the newborn baby needs its *mother*, the cultural milieu understood instead that 'the newborn baby needs its *parents*'. It was exactly at that time that the doctrine of the participation of the baby's father at birth developed. This is how we jumped directly from a generation that had no idea of what the interaction between mother and newborn baby can be, to a generation that had no idea of what a birth can be when, for example, there is nobody around the labouring woman apart from an experienced and silent midwife.

In spite of such cultural hurdles there are reasons for optimism. If, during the twentieth century, it has been possible to discover the basic needs of newborn babies, why – thanks to twenty-first-century physiology – can we not rediscover the basic needs of labouring women? We are not in the realm of utopia.

CHAPTER ELEVEN

Avenues for research

Having clarified what the basic needs of newborn babies are, scientists have opened a period of transition during which we are waiting for a clear understanding of the basic needs of labouring women. Ideally the newborn baby needs to interact with a mother who has given birth by herself and is therefore in a specific physiological state. This is why the needs of newborn babies are rarely ideally satisfied in our societies, since most women need pharmacological or operative assistance. Is it possible to challenge the cultural conditioning related to childbirth?

We already have at our disposal physiological concepts that are promising avenues for research. Most of these physiological concepts are already well established. The point is that we have to study them in more depth and, above all, to 'digest' them. In the case of a physiological process that is highly altered and repressed by the cultural milieu, the real difficulty is not acquiring knowledge, but digesting the knowledge provided by scientific perspectives.

A basic simple physiological concept

The concept of adrenaline-oxytocin antagonism is a typical example of a simple established concept that is not well 'digested'. It means that when mammals release adrenaline they cannot release oxytocin. Oxytocin has been presented by Kerstin Uvnäs Moberg as 'the mirror image of adrenaline'.[1] We use the word 'adrenaline' as a simplified way of referring to the 'fight and flight system'. We should say 'hormones of the adrenaline family'.

Such an antagonism has been understood for a long time and evaluated in relation to the response of the uterine muscle[2] and to the milk ejection reflex.[3] The first data have been confirmed by clinical studies in humans.[4,5] It is significant that among the drugs used to stop uterine contractions there are the so-called beta-mimetic drugs: their targets are uterine cells sensitive to adrenaline ('beta-receptors'). These drugs, by imitating locally the effects of adrenaline, have a weakening effect on uterine contractions. It is well known that adrenaline is a hormone that mammals, including human mammals, release in emergency situations, particularly when they are scared, when they feel observed, or when they are cold. The practical conclusion is that to give birth a woman needs to feel secure, without feeling observed, and be in a warm enough place.

It is apparently simple. However, many books for the general public provide recommendations illustrating the frequent difficulties in fully understanding basic concepts. For example, the common advice to be upright and to walk during labour, based on the simplistic idea that gravity can facilitate the descent of the baby, does not take into account that a prerequisite for the labour to establish itself properly is a low level of adrenaline. Let us imagine a woman in early labour who is passive, for example lying down on one side. Such a situation should simply be interpreted as an indication of a low level of adrenaline compatible with an easy progress of labour. To interfere by encouraging this woman to stand up and walk is at best useless. It is probably counter-productive. It can be unpleasant. An analogy with a situation always associated with a low level of adrenaline can help in understanding how unpleasant it can be: let us imagine that at the very time when we fall asleep we hear a voice saying 'get up and walk'.

There is a common assumption among the natural childbirth movements that labouring women need energy, like, for example, marathon runners. Women are advised to consume supplements of sugar (pieces of sugar, glucose tablets, soft drinks, etc.). This advice is also based on difficulties in assimilating what we know about oxytocin release in relation to the levels of adrenaline; in other words about the prerequisites for the labour to establish itself properly. When all the voluntary muscles are at rest the need for glucose is dramatically reduced. The observations by Paterson are highly significant.[6] They date back to the 1960s, when caesarean sections were routinely performed under general anaesthesia. In order to explain their importance, we must recall that when there are ketone bodies in the urine, it simply means that fatty acids have been used as a fuel, because there was a shortage of glucose. They found that ketone levels were higher in women who had been starved for twelve hours before an elective caesarean under general anaesthesia than they were for women who had been in labour. This confirms that labouring women spend less energy than those who are waiting for an operation *without* being in labour.

Furthermore, there is an abundant medical literature evaluating the dangers of giving simple sugars to labouring women. Risks for the newborn baby include severe hypoglycemia (low blood sugar level) and jaundice.[7,8,9,10] It is easy to understand that simple sugars can easily reach the foetus via the placenta, and that the excessive insulin production generated by the baby's pancreas can induce dangerous fluctuations of its glycaemia and disturb its liver metabolism. This is why, in most hospitals, when an intravenous drip is used as a vehicle for drugs, it is not usually a glucose solution. However, particularly in the case of out-of-hospital births, it is still common to talk about the need for energy.

The concept of neocortical inhibition

While the concept of adrenaline-oxytocin antagonism is simple and universal among mammals, the concept of neocortical inhibition offers an opportunity to refer to the particularities and the complexity of *Homo sapiens*. To analyse this concept we must once more present a human being as a member of the chimpanzee family that has developed to an extreme degree the part of the brain called the neocortex. We have already mentioned that giving birth may be considered an involuntary process that is under the control of archaic brain structures responsible for the release of a specific hormonal flow. The point is that the activity of the powerful 'new brain' can inhibit the functions of older brain structures. This is the meaning of 'neocortical inhibition', a fruitful search term when exploring medical and scientific databases. This is one of the reasons why human births are comparatively difficult. It is essential to understand the solution Nature found to overcome such a human handicap. It is simply that during the birth process the neocortex must stop working. Giving birth is not the business of the brain of the intellect. When our neocortex is at rest we have more similarities with other mammals.

Even in the twenty-first century, there are still some rare people who can understand from experience this solution that Nature found to overcome specifically human difficulties. They know that during an easy, unmedicated birth, there is a time when the labouring woman behaves as if she is cutting herself off from our world. She becomes indifferent to what is happening around her. She tends to forget what she has read, what she has learned and what her plans were. She may behave in a way that would usually be considered unacceptable in a civilized woman: for example, she may dare to scream or to swear. She may be impolite. She may talk nonsense. She may find herself in the most bizarre, unexpected posture. These postures are often primitive, quadrupedal. When the labouring woman is as if 'on another planet', it means that there is no neocortical control anymore.

Interpreting such a reduced neocortical control is a key for understanding the universal needs of labouring women. One can easily express the leading point in one short sentence: the birth process needs to be protected against all stimulants of the neocortex. Once more it is apparently simple, since the main stimulants of the human neocortex have been identified. Theoretically, it should be easy to eliminate these stimulants. In reality thousands of years of socialization of childbirth make this extremely difficult.

The first difficulty is about the concept of protection against the effects of language, the main stimulant of the human neocortex. Language is probably the most widespread negative interference during the birth process. In other words, silence appears a basic need. Even if this is theoretically simple, it cannot be accepted overnight, since it is challenging our cultural conditioning. Interestingly, we can learn from comparisons of birth statistics in 'talkative cultures' (e.g. Southern Italy and other Latin cultures) and, on the other hand, cultures with a high capacity for silence (e.g. Scandinavian countries and Japan). Meanwhile, to face these difficulties, birth attendants must be gradual in training themselves to avoid the use of language during labour. They must start by trying to eliminate the kind of language that is particularly powerful in terms of neocortical stimulation. This includes any rational language: let us think, for example, of the effects of talking about centimetres in front of a labouring woman. It also includes language expressing a question. When we are asked a question, we activate our neocortex in order to prepare an answer. How to explain this simple fact and its implications? When scientific knowledge and repressed common sense are in conflict with cultural conditioning, an analogy is often the most effective way to go the step further. The analogy with sexual intercourse is particularly explicit. Imagine a couple in a pre-orgasmic state. Suddenly the woman asks her partner: 'What do you want for dinner?'. Through neocortical stimulation, this simple question can interfere with physiological processes.

Light is another stimulant of the neocortex. Until recently, this was empirical knowledge: the figurative meanings of words such as 'illumination', 'enlightenment' and 'brightness', and of idioms such as 'in the light of' or 'to throw light on' attest a deep-rooted understanding of the effects of light on cognitive functions. Today the effects of light on neocortical activity have been demonstrated, particularly by neurophysiologists and electroencephalographists. Furthermore, one can now offer physiological interpretations since we know about the 'darkness hormone'. In general we switch off the lights before going to bed as a way to facilitate the release of this darkness hormone called melatonin. One of the properties of melatonin is to reduce neocortical activity. This is why it is usually easier to fall asleep in the dark and why, in a birthing place, there is probably a difference between a dim light and a bright light. For those who have digested the concept of neocortical inhibition, this is an inescapable research topic at a time when there are reasons to improve our understanding of the birth process. It is a crucial topic in the age of electricity. Interestingly, apart from *Birth Without Violence*, by Frédérick Leboyer, most classical books about childbirth ignore this subject.

One must also consider situations that are associated with an increased neocortical activity. This is the case of what happens when one feels observed. When one feels observed, we have a tendency to observe ourselves. We pay attention to ourselves. This is a way to interpret privacy as a basic need during labour. This need is not usually taken into consideration at a cultural level. It has to be discovered by health professionals and the general public as well. Let us recall the current epidemic of videos and photos of so-called natural childbirth: there are almost always at least two or three people watching the labouring woman … plus a camera.

The perception of a possible danger is another situation that implies the need to pay attention. It is therefore associated with a stimulation of the neocortex. This is one way to conclude that to feel secure is a basic need.

After taking as a point of departure the two well-established concepts of adrenaline-oxytocin antagonism and neocortical inhibition, it appears that the physiological perspective always leads to very simple conclusions. It is as if today we need scientific disciplines to unveil a common sense that has been repressed for thousands of years by cultural milieus.

CHAPTER TWELVE

Repressed common sense

When contrasting our cultural conditioning with the lessons of modern physiology, we have been constantly repeating that today we need scientific perspectives to unveil common sense. We gave the typical example of how the basic needs of a newborn baby have been discovered during the second half of the twentieth century: it appeared from an accumulation of data that the newborn baby needs its mother. This was challenging thousands of years of routine separation of mother and newborn baby with delayed initiation of breastfeeding. At a critical time in the history of childbirth, which is a critical time in the history of mankind, we must realize that many aspects of common sense have long been deeply repressed, particularly in the fields of reproduction and sexuality. We must also realize the power of emerging scientific perspectives: today it is common sense to claim that a newborn baby needs its mother.

If...

When mentioning two physiological concepts as avenues for research, we have reached extremely simple conclusions: ideally a labouring woman needs to feel secure, without feeling observed, in a warm and silent place. In reality it will probably take a long time for our cultural milieu to take on board such conclusions.

If our cultural milieus had had a deep understanding of 'the need to feel secure without feeling observed', the midwife would not have disappeared altogether in several parts of the world, or become a 'mini-doctor' following protocols, or just one of the members of a medical team, which is what has happened elsewhere. The physiological perspective is suggesting questions about the prototype of the protective person. These questions just need to be phrased the right way to make the answers obvious. In an ideal world, the prototype of the person with whom one feels secure without feeling observed is the mother. Today we need the physiological perspective to explain the specific role of the midwife as a mother figure. When the reasons for midwives are explained in modern scientific language, the translation into the language of common sense is easy.

If our cultural milieu had had a better understanding of the need to feel secure, there would not be the promoters of home births on the one hand and the promoters of hospital births on the other hand. It would be better understood that there is no universal recipe for feeling secure when giving birth. While some women feel more secure in a familiar place close to an experienced midwife perceived as a mother figure, others are conditioned in such a way that they need a modern environment with electronic beeps. If our cultural milieu had had a better understanding of the need to feel secure, the theoreticians of the twentieth century would have been more cautious before introducing the doctrine of the routine participation of the baby's father at birth. It is possible that one day imaginative scientists will develop research protocols demonstrating that, in general, a labouring woman feels more secure in the presence of a woman who has personal experience of giving birth than in the presence of a man who is out of context. Once more a facet of common sense will have been unveiled. Then it will be acceptable to claim that labouring women don't need 'emotional support': they need to feel secure.[1]

If our cultural milieu had a better understanding of the possible effects of language on the involuntary birth process, midwives would be trained to avoid asking questions of labouring women. If the physiological perspectives were more influential, those whose job is to select candidates for midwifery schools would be much more keen to find women of peaceful natures who can easily play the role of mother figure. Today most midwifery schools prioritise other criteria.

Analysing a concrete scenario

In order to illustrate the capacity of modern physiology to rediscover simplicity and to unveil repressed common sense, consider a situation I am familiar with. It is usually associated with an easy birth. This situation is almost unknown in our societies. We might even claim that it is at the limit of what is culturally acceptable. Imagine a woman in labour in a warm, dimly-lit room with nobody around, apart from one experienced and silent midwife sitting in a corner knitting. It is apparently simple. Every component of this scenario can be interpreted in the light of modern physiology.

It is easy to interpret the effects of a warm and dimly-lit environment as a way to reduce the level of adrenaline and to moderate neocortical stimulation. The fact that there is only one midwife must be emphasized. A woman in labour is more likely to feel observed if there are two midwives. This has been taken into consideration in societies as far apart as Persia, Hungary and South America. According to a Persian proverb, 'When there are two midwives, the baby's head is crooked'.[2] According to the Hungarian variant, 'Among many midwives the baby gets lost'. The Brazilian anthropologist Martha Azevedo told me about another variant of the same proverb among the Guarani Mbya from Angra. We described the midwife as 'experienced' and 'silent'. It is obvious that the woman in labour feels more secure (lower level of adrenaline) with an experienced midwife, while she is protected against the effects of language on neocortical activity if the midwife remains silent. Let us add that the midwife is not positioned in front of the labouring woman, watching her. She is sitting in a corner. It is not a relationship of observer and observed; this is further emphasized by the fact that the midwife is doing something else – knitting.

It is worth commenting on the knitting midwife.[3] When I was in the maternity unit of a Paris hospital, in 1953, midwives spent their lives knitting. I have heard modern midwives mocking this traditional attitude, but it needs to be reinterpreted in the scientific context of the twenty-first century. At the April 2004 British Psychological Society conference, Dr Emily Holmes, from Cambridge University, presented her studies of the effects of repetitive tasks, such as knitting, in stressful situations. In one of them volunteers were recruited to watch a video of real footage of car crashes showing dead bodies and a lot of blood. Some participants were given a repetitive task, such as tapping out a complex five-key sequence of numbers on a keypad while they watched. Those who were given such a task experienced fewer flashbacks over the following days than the others. The author concluded from her studies that repetitive tasks are an extremely effective means of reducing the levels of stress hormones. She also referred to the use of worry beads in many cultures, such as in Greece, as a way of coping with stressful situations. We are thus in a position to understand that when midwives spend hours and hours knitting, their own levels of adrenaline are kept as low as possible. To add further importance to these findings, we can now demonstrate, through sophisticated scientific methods, that emotional states are extremely contagious. This is thanks to the exploration of the 'Mirror Neuron System', an emerging scientific discipline using techniques of brain imaging. In such a scientific context, we can conclude that a birth attendant should take care to keep her own level of adrenaline as low as possible, since it is contagious.

When analysing the components of such a birth environment, my objective has not been to promote a model. It has been to reflect on common sense: what today is culturally unacceptable might in the future be accepted as common sense... thanks to scientific perspectives.

The science-common sense collusion

After referring to the power of modern science to unveil many aspects of common sense that have been repressed for thousands of years, we must also mention how more recent attacks on common sense can be easily denounced in the light of recent scientific advances.

We'll take as an example the recommendation to drink a lot when giving birth: 'don't get dehydrated'. I have heard of fathers who had been trained to remind their partner every so often to have a drink. A full bladder is the immediate obvious effect of this fashionable recommendation. Once more we need first the point of view of physiologists. According to their perspective, oxytocin and vasopressin, hormones released by the posterior pituitary gland during labour, are water retention hormones, to the point that vasopressin is usually called the 'antidiuretic hormone'. This is also the case with oxytocin.[4] So, theoretically, there is no risk of dehydration when under the effects of such hormones. It is as if Nature had found a neat way of keeping the bladder empty during labour. Interestingly, mammals don't drink when giving birth. Of course, if a woman suddenly needs a sip of water before the foetus ejection reflex, she can have it. It is also well understood that an excess of liquid can make the uterine contractions weaker. Before the advent of tocolytic drugs (substances used to make the uterus quiet), water intake had been studied to try to delay premature births. This is easily interpreted by physiologists: it makes sense that an excess of liquid makes water retention hormones less useful and therefore inhibits the release of oxytocin and vasopressin.

The physiological perspective is supported by clinical

observation. One cannot find, in the medical literature, any studies about dehydration during labour. On the other hand, the risks of water intoxication and hyponatraemia (low blood level of sodium) are well-documented.[5,6]

An authoritative medical journal has published review articles about water intake.[7] The overall conclusions are simple: if we are not thirsty, we don't need to drink. What a collusion between science and common sense![7]

CHAPTER THIRTEEN

The story is not finished

The story is not finished when the baby is born. Between the birth of the baby and the delivery of the placenta there is the so-called third stage of labour. We have already looked at this short phase when referring to what we presented as an important scientific discovery of the second half of the twentieth century. We could conclude that the newborn baby needs its mother and we contrasted the scientific point of view with our cultural conditioning and the effects of thousands of years of routine mother-neonate separation. We must add an essential clarification. One cannot consider the basic needs of the newborn baby from the viewpoint of physiologists without keeping in mind the unprecedented particularities of childbirth during the twenty-first century. This viewpoint implies that the mother has given birth vaginally and is in a specific hormonal balance following the birth process. In fact this is not a common situation in the age of pharmacological assistance and simplified caesareans. We are in a transitory phase of history when the basic needs of newborn babies are already well understood, but we still have to rediscover the basic needs of labouring women. This is why our priority has been to focus on established physiological concepts that can be presented as promising ways to improve our understanding of the birth process.

The mother still has needs just after the birth of the baby. The team headed by Kerstin Uvnäs Moberg, in Sweden, has demonstrated that just after giving birth a mother has the capacity to reach a level of oxytocin that is higher than for the delivery itself.[1] This peak is vital since it is necessary for a safe delivery of the placenta with minimal blood loss, and also because oxytocin is the main component of the cocktail of love hormones a woman is supposed to release during that phase of labour. Knowing that oxytocin release is highly dependent on environmental factors, we must wonder what kind of environment can influence this special hormonal peak just after the birth of the baby. The first condition is that the mother is not cold. Regina Lederman found that the level of adrenaline can return to normal as early as three minutes after birth.[2] She has therefore demonstrated how crucial this short period of time is, and confirmed what can be learned from clinical observation. When asked what to prepare for a home birth, I only talk about electric radiators and extension cords, so that warm blankets or towels are constantly available. If a woman is shivering just after the birth of the baby, it simply means that she is not warm enough.

The second condition is that the mother is not distracted when discovering her baby. The mother needs to feel the contact with the baby's skin, to look at the baby's eyes, and to smell the odour of her baby. Any distraction can bring her back down to earth and inhibit the oxytocin release. It would take volumes to mention all the possible ways in which mothers are distracted just after the birth. Some of these distractions are features of our societies and preventable; for example, a baby is just born and a telephone rings. Other distractions are related to tradition. Imagine a woman who has just given birth. She is discovering her baby. She has forgotten the rest of the world. Then, a birth attendant comes with instruments to cut the umbilical cord. It is a powerful distraction. The mother will come back down to earth. She will not reach her vital peak of oxytocin. Meanwhile, before one understands the concept of protection against distractions, we must understand and accept the widespread recommendation to rely on a substitute for natural oxytocin to facilitate the delivery of the placenta. This is currently the best way to save lives. It is particularly important in developing countries where a specific drug (misoprostol) is cheap, can be stored at room temperature, and does not need to be injected. It takes ten minutes to learn how to use misoprostol. It will take decades to reverse thousands of years of deep-rooted cultural conditioning and to understand that at that stage the mother needs the baby. On the day when this interaction between mother and newborn baby is studied in depth and when it is scientifically confirmed that the mother needs her baby, a new chapter of the science-common sense collusion will have begun.

When considering this interaction between mother and baby we must constantly think in terms of hormonal balance. Oxytocin, in particular, is never released in isolation. That is why love has so many facets. In the particular case of the hour following birth, in physiological conditions, the high peak of oxytocin is associated with a high level of prolactin, which is the 'motherhood hormone'. This is the most effective situation for inducing love of babies. Oxytocin and prolactin complement each other. Furthermore, oestrogens activate the oxytocin and prolactin receptors. Today we are in a position to understand that all the different hormones released by mother and foetus during the first and second stages of labour are still circulating during the hour following birth; they have not yet been eliminated. All of them have a specific role to play in the mother-newborn interaction. The way oxytocin is released is a new avenue for research. In order to be effective, this release must be pulsatile: the higher the frequency of pulses, the more effective the hormone is.[3]

The maternal release of morphine-like hormones (endorphins) during labour and delivery is now well documented. We also now understand that the baby releases its own endorphins during the birth process.[4,5] There is no doubt that, for a certain period following birth, both mother and baby are impregnated with opiates. The property of opiates to induce states of dependency is well known, so it is easy to anticipate how the beginning of a 'dependency' – or attachment – is likely to form.

Even hormones of the adrenaline family (often seen as hormones of aggression) have an obvious role to play in the interaction between mother and baby immediately after birth. During the very last contractions before birth the level of these hormones in the mother peaks. That is why, in physiological conditions, as soon as the 'foetus ejection reflex' begins, women tend to be upright, full of energy, with a sudden need to grasp something or someone. One of the effects of such adrenaline release is that the mother is alert when the baby is born. Think of mammals in the wild, and we can more clearly understand how advantageous it is for the mother to have enough energy – and aggression – to protect her newborn baby if need be. Aggression is an aspect of maternal love. It is also well known that the baby has its own survival mechanisms during the last strong expulsive contractions, and releases its own hormones of the adrenaline family. A rush of noradrenaline enables the foetus to adapt to the physiological oxygen deprivation specific to this stage of delivery. The visible effect of this hormonal release is that the baby is alert at birth, with eyes wide open and dilated pupils. Human mothers are fascinated and delighted by the gaze of their newborn babies. It is as if the baby were giving a signal, and it certainly seems that this human eye-to-eye contact is an important feature of the beginning of the mother and baby relationship in humans. The highly complex role of hormones of the adrenaline-noradrenaline family in the interaction between mother and baby is a new area of study. A small number of animal experiments have paved the way for further research. Mice that lack a gene responsible for the production of noradrenaline leave their pups scattered, unclean and unfed - unless they are injected with a noradrenaline-producing drug when giving birth.

Our current knowledge of the behavioural effects of different hormones involved in the birth process helps us to interpret the concept of a critical period introduced by ethologists. It is clear that all hormones released by the mother and by the baby during labour and delivery are not eliminated immediately. It makes sense that all of them have a specific role to play in the interactions between mother and baby.

CHAPTER FOURTEEN

Labour pain revisited

Not only has modern science the power to challenge deep-rooted cultural conditioning, but it can also provide keys for interpreting ancient messages. It is as if valuable messages about human nature are likely to disseminate and to survive over millennia whatever the cultural conditioning. We are reaching a phase of history when it is possible to decode some of them.

The best example of such an unexpected power of modern physiology is related to the association of the consumption of the fruit of the tree of knowledge (Genesis 2,17) with difficult and painful labour (Genesis 3,16). The meaning of such an unnoticed association becomes suddenly clear in the light of the concept of neocortical inhibition: the activity of a highly developed neocortex can inhibit the involuntary process of birth.

Some millennia later, a legendary mother sent a complementary message. When in labour, she found a way to overcome her human handicap and to reduce her neocortical activity. She accepted her mammalian condition. She humbly gave birth in a stable, among other mammals. Thanks to this birth outside the human community, her baby was protected against invasive cultural interferences.

Interestingly, his name has remained associated with love.

A protective physiological system

In the current scientific context, a reduction in neocortical activity appears as one of the components of a physiological system of protection against labour pain.

There has recently been the beginning of a paradigm shift in our interpretation of labour pain. Until now, according to the dominant ways of thinking, the pain could be eliminated while the rest of the physiological processes could be maintained. Today we are in a position to understand that there is physiological pain during labour, but that there is also a physiological system of protection against pain. The important point is that the components of the physiological system of protection against pain have several roles to play, besides pain relief. In other words the pain is part and parcel of the physiological process: one cannot electively eliminate the pain without altering other aspects of the physiological process.

We have known since the late 1970s that mammals in general and women in particular control the pain of labour by releasing morphine-like substances commonly called endorphins.[1,2] This release of endorphins is one of the components of the protective system against pain. We learned at the same time that these endorphins (beta-endorphins) stimulate the secretion of prolactin, the motherhood hormone and the key hormone of lactation.[3] It is therefore possible today to interpret a chain of events that starts with the physiological pain of labour and leads to the release of a hormone considered necessary for the secretion of milk. Any attempt to electively eliminate the pain will neutralize the whole chain of events.

We can make similar comments about a reduction of neocortical activity as another component of this physiological system. This aspect of birth physiology, which we presented as a condition for the release of a specific complex hormonal flow, can also be presented as a component of the protective system. When the neocortex is at rest the pain is not integrated into the central nervous system in the same way as in other situations. This is how we can explain women screaming 'it hurts' while also being 'on another planet'. In the days that follow they claim that the birth was not painful. The depression of memory has obvious protective effects. It is an effect of a reduced neocortical activity, and also of the well-known properties of opiates in general, and therefore of endorphins. Furthermore, the amnesic effect of oxytocin has been demonstrated among humans, through experiments confirming the results of animal studies: memory tests have been done after a single dose of intranasal oxytocin.[4]

After such a paradigm shift, the primary objective should not be to make births painless. The primary objective should be phrased differently. It should be to make births as *easy* as possible, so that the need for pharmacological assistance is reduced. When a birth is fast and easy it implies that the hormonal balance is appropriate and that the pain is controlled by the physiological protective system. Finally, once more, we come to the conclusion that the basic needs of labouring women must be rediscovered.

Meanwhile

Meanwhile, most modern women have difficult births, including many of those who had planned to stay at home. The pain can reach a pathological degree, with such high levels of stress hormones, particularly adrenaline, that the pain itself is the obstacle to the progress of labour. Today such situations are usually treated by epidural anaesthesia. This may be effective at breaking a vicious circle, so that dilation can progress, but it is often also associated with a drip of synthetic oxytocin and the risk of the birth culminating in a vaginal operative delivery (ventouse or forceps) is high.

In the 1970s I developed an interest in the particular case of labours associated with intense pain in the lower back in which dilation, although already well advanced, cannot progress further. I was looking at non-pharmacological ways of breaking the pathological vicious circle. This is how I introduced in the medical literature the use of intracutaneous injections of sterile water[4] and birthing pools.[5]

Today an accumulation of published hard data confirms what the priority should be in the age of simplified techniques of caesarean section. Extreme caution regarding the pharmacological approach should be the watchword. Pharmacological assistance is, more often than not, associated with difficult births by the vaginal route. All the components of medically assisted vaginal births have documented side-effects that have been underestimated until recently.

For example, it appears that epidural anaesthesia tends to increase the risk of operative delivery by interfering with the process of rotation.[6] The most common technique of epidural anaesthesia (with an opioid analgesic) has documented negative effects on the quality and the duration of breastfeeding.[7] Paradoxically, although it is the most common medical intervention in childbirth, the side-effects of synthetic oxytocin have not been extensively studied. However, there are theoretical and even epidemiological reasons to suggest that synthetic oxytocin (syntocinon or Pitocin) should be used with caution. It probably interferes with the neurodevelopment of the child, as confirmed by a study of risk factors for Attention Deficit Hyperactivity Disorder,[8] and by the Malaga study.[9] At the University Hospital of Malaga, Spain, 400 birth records dated 2006 were analysed and the doses of synthetic oxytocin received by the mothers during labour were evaluated. Then the families were interviewed in 2011, while the children were assessed through the Batelle developmental inventory (with a minimum of ninety-five items tested). According to non-statistically significant preliminary findings, further studies are necessary to evaluate, in particular, the effects on psychomotor development. For children exposed to synthetic oxytocin the chances of an abnormal result were 8.9 times higher after vaginal birth than after caesarean. Furthermore, this study could demonstrate a significant negative effect of the use of synthetic oxytocin on the duration of breastfeeding, with a dose-effect response.

In fact there is usually such a complex combination of obstetrical interventions that epidemiologists have difficulty designing research protocols that will detect the specific side-effects of one particular intervention. For the same reason, the results of many studies are difficult to interpret. For example, epidural anaesthesia is often associated with synthetic oxytocin, while operative deliveries by the vaginal route (forceps or ventouse) and emergency caesarean sections are often preceded by pharmacological assistance.

This is why we must also rely on studies turned up by searches for keywords such as 'asphyxia at birth'. In practice asphyxia may be considered a proxy for difficult birth by the vaginal route or last-minute emergency caesarean section after a long phase of synthetic oxytocin. A large study from the Stockholm county demonstrated a strong association between signs of asphyxia at birth and schizophrenia after adjustment for many confounding factors.[10] According to another Swedish study, suicides involving asphyxiation were closely associated with asphyxiation at birth,[11] while according to a Danish study, asphyxia at birth was a risk factor for severe handicap at age four.[12]

There is food for thought in the results of an enormous study in eighteen counties and three cities in China. In that part of China the overall rate of caesarean section was 56%. The objective was to examine the association between mode of delivery and childhood psychopathology. More than 4,000 children (all of them first babies aged four to six) were assessed with the Child Behaviour Checklist, an instrument to evaluate a child's emotional (internalising) and behavioural (externalising) problems. Differences scores were analyzed among three groups of children: one group born by pre-labour caesarean section on maternal request, another one born by vaginal delivery with forceps or ventouse, and another one born vaginally without forceps or ventouse. It appeared that the likelihood of childhood psychopathological problems was the lowest in children born by pre-labour caesarean on maternal request, followed by those born by the vaginal route without forceps or ventouse, and the highest probability was observed in those born with forceps or ventouse.[13] Once more, the conclusion is that the priority is to avoid difficult births by the vaginal route.

These conclusions should be considered alongside the results of a British study of the response to vaccination of eight-week old babies in relation to how they were born.[14] One group was born vaginally without forceps or ventouse. Another group was born vaginally with forceps or ventouse, while a third group was born by elective pre-labour caesarean section. Two criteria were used to evaluate the stress responses: measuring the duration of crying with a stopwatch and measuring saliva cortisol values just before and twenty minutes after the injection. Whatever the criteria, the greatest response was shown among babies born with forceps or ventouse and the least response in those born by pre-labour caesarean section. The results of this study provide another warning about the possible negative effects of difficult births by the vaginal route.

In considering the results of these two studies, remember that babies born by pre-labour caesarean section have not been exposed to synthetic oxytocin. It is highly probable that, in both the Chinese and the British context, this was not the case for most of the other babies in the studies. As usual, the use of synthetic oxytocin was not taken into account.

We must emphasize that two kinds of births have not been considered in these studies and, in general, are not considered in the medical literature: firstly, vaginal births without drips of synthetic oxytocin and other kinds of pharmacological assistance, and secondly in-labour non-emergency caesarean sections. Until now it has been commonplace to ignore the fact that an in-labour caesarean can be planned, and also that it can be decided on before a state of emergency is reached during a trial of labour. In the age of simplified techniques of caesarean section, and supposing that we can rediscover the basic needs of labouring women, the paradigm shift will be characterized by simplified strategies adaptable to most cases: a straightforward vaginal birth, or an in-labour non-emergency caesarean as the usual alternative. As we mentioned previously, there are reasons to avoid both pre-labour caesareans preceding foetal signals of readiness to be born, and last-minute emergency operations. We must add to the reasons to avoid pre-labour caesareans the results of preliminary bacteriological studies. It seems that, from a bacteriological perspective, there are significant differences between the milk of women who gave birth via pre-labour c-section compared with the milk of those who had an in-labour c-section. There are no detectable differences, on the other hand, between the milk of those who gave birth vaginally and the milk of those who had an in-labour c-section.[15] As we have already mentioned, such differences according to the timing of the operation are confirmed by a Canadian study of four-month-old babies.[16] It seems that the paramount importance of these issues is not yet recognized by the relevant scientific disciplines. Such a paradigm shift should open a new phase in the history of medical research.

CHAPTER FIFTEEN

No paradigm shift without language shift

I have repeatedly used the term 'in-labour non-emergency caesarean'. This is an example of the advent of a new vocabulary adapted to the paradigm shift which the power of modern biological sciences makes imaginable. Such a language shift would also imply that some terms become less useful.

Towards a new vocabulary

'Foetus ejection reflex' is another typical example of a phrase that might become easily understandable and more commonly used, should the paradigm shift occur. This term was coined by Niles Newton in the 1960s when she was studying the environmental factors that can disturb the birth process in mice.[1] Twenty years later, with her support,[2] I suggested that we save this concept from oblivion; I was convinced it could be a key to facilitating a radically new understanding of the process of human parturition.[3]

The basic difference between humans and mice is that we have developed a huge and powerful neocortex, which covers more archaic structures. When our neocortex is at rest we have more physiological similarities with mice.[4]

An authentic foetus ejection reflex is possible in humans. It takes place when a baby is born after a short series of irresistible and powerful contractions, which leave no room for voluntary movements. In such circumstances, it is obvious that the neocortex is at rest and no longer in control of archaic brain structures in charge of vital functions such as giving birth. This is the case, as we have already seen, when civilized women behave in a way which would usually be unacceptable, daring to shout, to swear, or to be rude, for example. They seem to cut themselves off from our world. They can talk nonsense. During a real foetus ejection reflex, women can find themselves in the most unexpected, bizarre, often mammalian, quadrupedal postures. They seem to be 'on another planet'. They can be in an ecstatic state. In typical situations, it seems that the real climax is reached when the mother, still on another planet, is discovering her newborn baby. A mother told me that when she first had eye-to-eye contact with her baby, she 'saw the whole universe' in the eyes of her infant.

It is easy to explain why the concept of foetus ejection reflex is not understood after thousands of years of socialization of childbirth. It is precisely when birth seems to be imminent that the birth attendant has a tendency to become even more intrusive. The foetus ejection reflex can be preceded by a sudden, explosive expression of a fear, with a frequent reference to death.[5] The woman may say: 'Am I going to die?' 'Kill me!' or 'Let me die...' Instead of keeping a low profile, the well-intentioned birth attendant usually interferes, at least with reassuring rational words. These rational words can interrupt progress towards the foetus ejection reflex. The reflex does not work if there is a birth attendant who behaves like a coach, or an observer, or a helper, or a guide, or a 'support person'.[6] It is exceptionally rare if the baby's father participates in the birth. The foetus ejection reflex can also be inhibited by vaginal examinations, eye-to-eye contact, or by the imposition of a change of environment, as would happen when a woman is transferred to a delivery room. It is inhibited when the intellect of the laboring woman is stimulated by any sort of rational language, for example if the birth attendant says: 'Now you are at complete dilation. It's time to push.' In other words, any interference tends to bring the labouring woman back down to earth, and tends to transform the foetus ejection reflex into a second stage of labour which involves voluntary movements.[7]

The deep-rooted belief that the presence of a specialized person is the basic need of labouring women and newborn babies is another reason why the birth climax cannot be understood. In our societies unassisted deliveries occur occasionally, by accident. This is more often than not in the case of fast births. Either the midwife could not arrive in time, or the mother could not reach the hospital in time. Because such births are easy, you might think that it would be an opportunity for some women to reach a real climax. It is not so in most cases, because of the cultural conditioning that a woman is unable to give birth by herself. For example, if the husband/partner is around, he is usually in a state of panic wondering what he should do, who will 'deliver the baby' and who will cut the cord.

Before the possible paradigm shift, those who understand the foetus ejection reflex realize how useless it is to exchange views with others on issues such as breech birth, posterior position of the baby's head, shoulder dystocia, brachial palsy, or perineal lacerations. All scientific studies published in the medical literature about the best way to 'manage' particular obstetric situations or particular phases of labour are conducted in environments where the foetus ejection reflex is ignored and inhibited. And no foetus ejection reflex occurs when the birth process is 'managed'.

After the possible paradigm shift, the art of midwifery should become the art of creating the conditions for a foetus ejection reflex.

Avoidable terms

We have presented the birth process as an involuntary process that needs to be protected. It makes sense to assume that this protective attitude should start before the birth, since it is well understood that the prerequisite for labour to establish itself properly is a low level of anxiety. In other words, the more a pregnant woman is subjected to anxiety-provoking stimuli, the more difficult the birth process may be. A low level of anxiety is also prerequisite for the optimal growth and development of the baby in the womb. The preliminary practical questions must therefore focus on what can influence the emotional state of pregnant women.

Since pregnancies today are highly medicalized, we must first analyse the vocabulary commonly used in the framework of prenatal care. One way to protect the emotional state of pregnant women is to avoid terms usually associated with a 'nocebo effect'. The nocebo effect is a negative effect on the emotional state of pregnant women and indirectly of their families. It occurs whenever a health professional does more harm than good by interfering with the imagination, the fantasy life or the beliefs of a patient or pregnant woman.

'Gestational diabetes' as a typical example

'Gestational diabetes' is a typical example of a term with a strong nocebo effect. It has the power to transform a happy pregnant woman into an anxious or depressed one. There are no symptoms of gestational diabetes: it has been called a 'diagnosis still looking for a disease'. It is the interpretation of a test (glucose tolerance test or GTT) that is routine in many countries: if the glycaemia (amount of glucose in the blood) is considered too high after absorption of sugar, the test is positive. One of the roles of the placenta is to manipulate maternal physiology for foetal benefit: the placenta, as an endocrine gland, is the advocate of the baby. At a certain phase of foetal development, there is an increased demand for glucose. Some women (according to their metabolic type) must make a bigger effort than others to satisfy this demand. These women are labelled as having 'gestational diabetes'. A huge study, involving all mothers and newborns registered by the Canadian Institute for Health Information from 1984 to 1996 (even-numbered years only) could not detect any beneficial effects of routine screening on pregnancy outcomes.[8] A review of the medical literature by the US Preventive Services task force reached similar conclusions.[9]

I have been given the opportunity, in a mainstream medical journal, to suggest that this term should be considered useless.[10] Instead of using it, it would be more cost-effective to routinely spend longer than usual discussing in depth with *all* pregnant women several aspects of their lifestyle, in particular the importance of daily physical activity and, in the age of soft drinks and white bread, issues such as high versus low glycaemic index foods. It is true that women who have been diagnosed as having gestational diabetes are more at risk than others of developing type-2 diabetes later on in life. This fact is used as an argument for routinely screening pregnant women, and medicalized antenatal care is seen as the 'opportunity of a lifetime' in detecting women at risk of becoming diabetic.[11] Would it not be better to make antenatal care the 'opportunity of a lifetime' for reconsidering several aspects of our modern lifestyle? Instead of focusing on the prevention of a limited number of maternal disorders, would it not be more advantageous to positively promote health and to develop long-term thinking?

One of the side-effects of the term 'gestational diabetes' is to transform the interpretation of the results of a test into a disease. The status of disease implies that complications have been identified. It is commonplace to claim that macrosomia (a big baby) is the main complication. This should be considered an *association*. It is obvious that the energy requirements of a big baby are not the same as the requirements of a small one: the mother, who must make a bigger effort than others, is labelled as having 'gestational diabetes'. The worst scenario is when the so-called disease is actively treated, usually by insulin. This attitude is based on a lack of understanding of the physiological processes. It is not understood that to satisfy the requirements of her baby the mother must increase her insulin resistance in order to maintain nutrient flow to the growing foetus. Active treatment will neutralize the demand expressed by the placenta and will inhibit the growth of the baby, particularly the growth of its brain. It is significant that active treatment tends to moderate birth weight, but does not influence Body Mass Index at age four- to five-years-old.[12] Children must catch up. From an exploration of the Primal Health Research Database using the the keyword 'catch-up growth', it appears that the need to catch up after intra-uterine deprivation is a handicap with negative long-term consequences.

The nocebo effect of the term 'gestational diabetes' is becoming a serious issue. The use of enlarged criteria to interpret the tests is one of the reasons why the number of women diagnosed with gestational diabetes is increasing.

Other examples

The strong nocebo effect of the word 'haemorrhage' is underestimated. This word is associated with the concepts of disaster and death. Many pregnant women have a low platelet count (platelets are tiny blood cells involved in coagulation). A low platelet count in pregnancy should not even be mentioned when it is not associated with a pregnancy disease such as pre-eclampsia, eclampsia or a chronic disease such as thrombopenic purpura. However, many pregnant women – particularly those who are planning a home birth – are warned about the risk of haemorrhage because they don't have a sufficient number of platelets. Many others are told that they are at risk of dangerous haemorrhage because their haemoglobin concentration is low: they must treat their 'anaemia'. In reality, in most cases, thanks to the demand expressed by an active placenta for an increased blood volume, the haemoglobin concentration is low because it is measured in a diluted blood: this is an advantage to compensate a blood loss. Other women are told about the risk of haemorrhage if they do not accept an injection of synthetic oxytocin (or another specific drug) to facilitate the delivery of the placenta. In practice the word haemorrhage is not useful in the framework of antenatal visits. For example, it is possible to talk about trying to reduce blood loss, instead of 'preventing a haemorrhage'.

The term 'hyperemesis gravidarum' has fascinated the media recently after a British princess was hospitalized early in her first pregnancy. When a pregnant woman is vomiting, there are two kinds of typical scenario, according to the personality of the practitioner. Some practitioners make a big fuss of the situation and it does not take long for them to establish the brilliant and impressive diagnosis of 'hyperemesis gravidarum'. Their immediate preoccupation is with how to treat the disease. This attitude can induce a cascade of unpredictable effects.

Another scenario is possible. Some practitioners, after expressing their compassion for such an unpleasant situation, use a different vocabulary. They take the time to recall that about two-thirds of pregnant women have morning sickness at the beginning of their pregnancy. They explain that this 'reduced digestive tolerance' is now considered protective.[13] It is a sign of positive hormonal changes induced by the pregnancy. It is well known that vomiting pregnant women are at very low risk of miscarriage and usually give birth to healthy, normal-weight babies.[14,15] This is an important point, because many vomiting pregnant women are in anguish about the growth of their baby, and because the symptoms of morning sickness can be amplified and prolonged by emotional factors.[16] Furthermore, it has been demonstrated that the risks of foetal abnormalities such as cleft lip[17] and anomalies of the penis (hypospadias)[18] are significantly reduced after pregnancies in which the mother has had episodes of vomiting. Interestingly, traditional farmers knew about the advantages of food restriction in early pregnancy. Just after conception and for a limited period of time, they put ewes in poor pastures.

After such a conversation between doctor and pregnant woman, the issue of treatment may appear secondary. Since, according to some statistics,[19] the incidence of real hyperemesis can be as low as 0.3%, it is possible, in most cases, to start with non-invasive treatments such as, for example, hypnosis or acupressure on the Neiguan point on the forearm.[20] Of course, it is another matter when real hyperemesis starts after the first trimester of pregnancy.[21]

I saw a photo of a gorgeous princess playing hockey. Some days later I saw a photo of the same princess giving an award to an Olympic champion. Between the two events, she had been hospitalized for 'hyperemesis gravidarum'...

CHAPTER SIXTEEN

Love as an evolutionary handicap

What are the limits of human adaptability? This is the kind of question inspired by an analysis of our cultural conditioning, particularly our conditioning related to childbirth and its recent reinforcement. We have used the term crisis. One cannot appreciate the nature and the amplitude of the current crisis without looking back at the previous spectacular crisis in the history of mankind, namely the Neolithic revolution.

Since *Homo sapiens* began to colonize all the continents, from about 100,000 years ago until recently, our ancestors obtained their food from wild plants and animals. They were taking advantage of what nature could offer. They were gathering, hunting, scavenging and fishing. Suddenly, around 10,000 years ago, *Homo sapiens* started to domesticate plants and animals in places as diverse and as far apart from each other as the Tigris and Euphrates valleys in the Near East, South Asia, Central Asia, and Central America. The advent of agriculture and animal husbandry opened the Neolithic phase in the history of mankind. From that time, the strategies for survival of human groups were based on the domination of nature. Domination of nature implies the capacity to destroy life. Agriculture and animal husbandry gave a renewed importance to the concepts of territory and property, and therefore new reasons for conflicts between human groups. To develop the potential for aggression and to interfere with the development of several facets of the capacity to love became prerequisites for a human group to be successful.

The concept of critical period in the light of anthropology

Starting from such considerations, we can present cultural anthropology as another discipline confirming that the perinatal period is critical in the formation of human beings, in particular in the development of the potential for love and, inversely, the potential for aggression. From the time when it became an unprecedented advantage to develop the potential for aggression, there has been natural selection of human cultures. Cultures that have amplified the difficulties of human births and adopted invasive beliefs and rituals in the perinatal periods have come to dominate the others.

Where childbirth is concerned, we learn more from the common points between cultural milieus than from the differences.

Giving birth before and after the Neolithic revolution

From what we know about pre-agricultural societies, we can claim that there was a time in the history of humanity when women used to isolate themselves when giving birth, like all mammals. This is confirmed by a great diversity of documents. We know, thanks to Melvin Konner,[1] about the 'solitary and unaided births' that take among the African hunters and gatherers !Kung San:

> A woman feels the initial stages of labour and makes no comment, leaves the village quietly when birth seems imminent, walks a few hundred yards, finds an area in the shade, clears it, arranges a soft bed of leaves, and gives birth while squatting or lying on her side – on her own.

According to Jean Pierre Hallet, the Efe Pygmies, who lived in Zaire's Ituri Forest, had no rituals or beliefs disturbing the birth process.[2] Living in the middle of a dense forest, they had no conflict with other human groups and had developed a strong ecological instinct, in particular an enormous respect for trees. The Eipos, in New Guinea, studied by Wulf Schiefenhovel as early as in the 1960s, were not perfectly representative of a typical pre-agricultural ethnic group, since they had gardens and pigs. However, the births were not socialized and, in films made discreetly by Wulf and his wife, women are seen giving birth in the bush, without any assistance, through authentic foetus ejection reflexes similar to those some modern women can occasionally experience in ideal situations.[3]

Daniel Everett, a Christian missionary and linguist, spent three decades studying and living with the Piraha, who live along the banks of the Maici River, in the Brazilian part of Amazonia. Their society, like that of the Eipos in New Guinea, could be described as transitional, since they have dogs and they cultivate manioc. However, they are mostly fishermen, gatherers and hunters and they have retained many features and social processes of pre-agricultural societies. This is what Everett wrote about childbirth[4]:

> When a Piraha woman gives birth, she may lie down in the shade near her field or wherever she happens to be and go into labor, very often by herself. In the dry season, when there are beaches along the Maici, the most common form of childbirth is for the woman to go alone, occasionally with a female relative, into the river up to her waist, then squat down and give birth, so that the baby is born directly into the river...

Once more, we see that women usually give birth unattended in pre-agricultural societies. This is all the more interesting because Everett had previously made comments on the issue of privacy. He had emphasized that in this ethnic group 'privacy is not a strong value'. However, the Piraha did need privacy to give birth and to make love, as if they knew that oxytocin has been called 'the shy hormone'. The information transmitted by Everett is important for other reasons. It is a valuable record of a society where many babies are born under water. Furthermore, Everett mentioned the case of an unattended breech birth that ended with the death of both mother and baby. When the labouring woman was heard continuously screaming, a male foreign visitor wanted to go to her and help her. He was told not to try. The attitude of local people (which may seem callous to our eyes, at first glance) was, however, based on a deep-rooted understanding of birth physiology. One cannot help an involuntary process. Any interference would have increased the risk of disaster.

The concept of a birth attendant is more recent than is commonly believed, although a mother or mother figure was possibly around when a woman was giving birth in primitive societies. The privacy of the birthing woman had to be protected against the presence of wandering men or animals.

After the Neolithic revolution, the common point between all cultural milieus has been the socialization of childbirth. It would take volumes to present a comprehensive study of the characteristics of a great number of cultures in relation to how they interfere.

In general, the degree of socialization of childbirth and the power of cultural interferences has gradually increased over millennia, with a spectacular acceleration during the twentieth century.

The protective mother or mother-figure became the midwife. This was an important step in the process of socialization. The midwife often played the role of a guide who dared to interfere using language. She became the one controlling the event, and also the agent of the cultural milieu in transmitting beliefs and rituals – using a great diversity of methods, including invasive procedures such as manual dilation of the cervix, compression of the abdomen, or traditional herbs. Another important step in the socialization of childbirth occurred when women started to give birth in the place where they spent their day-to-day life: home birth is comparatively recent in our history.

It is notable that although childbirth has been socialized for thousands of years, women always tended to protect the birthing place from the presence of men, particularly medical men. The role of doctors was limited to two spheres of competence. One was to intervene in desperate situations when the midwives called. Before the invention of forceps, usually all a medical man could do was to remove the infant piece-meal using hooks and perforators, or, if there was still hope of delivering a live child, to perform a caesarean section on the mother after her death. The realm of instruments is eminently male. The other sphere of competence of literate male physicians was writing about childbirth, mainly for the purpose of educating midwives and instructing other physicians on the supervision of birthing women. Since the medical man was called only for disasters, he had little opportunity to gain a real understanding of the birth process and the basic needs of labouring women. History helps us to interpret the deep-rooted and widespread lack of understanding of the birth process.

It is only in the middle of the twentieth century that the birth environment began to be 'masculinized'. The number of specialized doctors increased at lightning speed, and almost all were men. In around 1970 some women occasionally made a new demand (as a way of adapting to the 'industrialization of childbirth') for the participation of the baby's father at birth. Almost overnight it became a doctrine supported by theories. At the same time, sophisticated electronic machines invaded the delivery room, high technology being a male symbol. There was such indifference to the gradual masculinization of the birth environment that there were no serious discussions when midwifery schools began to accept male pupils. The almost total masculinization of birth had been achieved, as the ultimate result of the socialization of childbirth.

'Maternal urges' neutralized

In the current scientific context, one can claim that maternal love is the prototype of all facets of love.[5] One can add, through our current understanding of the physiology of 'maternal urges', particularly our knowledge of oestrogens, prolactin, oxytocin, and endorphins, that the strongest possible cocktail of love hormones a woman has the capacity to release during her whole life comes between the birth of a baby and the delivery of the placenta. It is highly significant that it is precisely during this phase of human life that, for thousands of years, successful cultural milieus (i.e. those one can study, the others having disappeared) have introduced the most invasive beliefs and rituals. We have already mentioned, using a different language, the universal neutralization of the 'maternal protective-aggressive instinct', without giving any definition of this instinct: to understand its nature, one just needs to imagine what would happen if one tried to pick up the baby of a mother gorilla who had just given birth. This routine neutralization of 'maternal urges' at a crucial phase of human life is so universal that we dare to present it as one of the bases of our civilizations.

Perinatal beliefs and rituals are apparently so diverse that it is easy to miss their common points: all of them disturb the interaction between mother and newborn baby and postpone the initiation of breastfeeding. Attitudes to early colostrum and the rituals of early cord-cutting are just examples that we have put forward because they have been documented on the five continents. Without trying to establish an exhaustive list we can offer other typical examples in order to bring to light their common points in spite of an apparent diversity.

What is striking is that several variants of the same rituals have often been documented in places separated by oceans. For example, there are many similarities between the birth rituals among the Arapesh from New Guinea, as reported by Margaret Mead,[6] and birth rituals in South America. Among the Arapesh, when a woman is giving birth, the father waits within earshot until the baby's sex is determined, when the midwife calls it out to him. To this information the father answers laconically: 'Wash it', or 'Do not wash it', which means that the child is or is not to be brought up. This clearly indicates that the baby is in the hands of the midwife, who is herself at the service of the father. Among the Myky from Mato Grosso, in Brazil, the mother is not authorized to touch the baby until the spiritual leader has confirmed that this particular baby should survive. Among ethnic groups in Amazonia, it is only after receiving permission from the godfather, who will be wearing his ceremonial clothes, that the mother can take care of her baby.

Dipping the baby in water at the very time when it might express the need to find the nipple is another example of a ritual documented in a great diversity of ethnic groups:[7] among American Indians (Sioux, Crows and Creeks), in Bolivia, in South India, in the Andaman Islands of the Bay of Bengal, or in Russia, for example. The accepted justification for most of these rituals is to make the baby strong. This is also the case, for example, of smoking the baby among the aborigines of Australia, or opening the doors in cold countries. We'll notice that one of the additional effects of all these practices is to distract the mother while she is in a specific physiological state that tends to direct her attention exclusively towards the baby. The distracting power of some birth rituals is particularly obvious. According to Sobonfu Somé, 'the keeper of rituals', when a woman is in labour in the Tagara tribe in Burkina Faso, the young children of the community wait nearby; as soon as they hear the baby's first cry, they all rush to the birthplace, shouting to 'welcome' the baby.[8] What a powerful way to abruptly interrupt the vital peak of oxytocin! What a powerful way to neutralize the 'maternal urges'!

The repression of maternal behaviour in the perinatal period must be looked at in the context of the 'Neolithic cataclysm'. In fact, all aspects of lifestyle have been turned upside-down, particularly all aspects of human reproductive/

sexual life. The prime mover in this shift was probably the advent of animal husbandry, with its variants of nomadic pastoralism and pastoral farming. Men started to realize, and even to overestimate, the role of males in reproduction. This is probably how fathers began to hold authority over women and children. There was a shift towards patriarchy, a social system in which the male is the primary authority figure. The universal transcendent emotional states were channelled towards the belief in one God as a paternal figure. Goddess religions and nature religions were eclipsed.

The control of all episodes of human reproductive/sexual life must be understood in the framework of the domination of nature. Some aspects of this domination are now being reconsidered. Where men-women relationships and genital sexuality are concerned, the pendulum is obviously swinging back. It is not yet the case where childbirth and, indirectly, breastfeeding are concerned.

CHAPTER SEVENTEEN

Reasonable pessimism

If you say that someone is emotional, it is pejorative.

If you say that someone is superintelligent without emotion, it is also pejorative.

If you talk about a well-balanced person, this is laudatory.

What is in the balance?

Whatever approach one might take towards understanding the human phenomenon, it must always take into consideration some aspects of the relationship between our two brains – the old one and the new one. The old one is ancient in the framework of the evolutionary process, since we share it with all other mammals. It is also old in the sense that it reaches maturity very early in our lives, during what we call the primal period.[1] It cannot be dissociated from the hormonal system and the immune system, with which it forms the 'primal adaptive system' as a complex network. The archaic brain structures govern instinctive behaviours and the basic mammalian emotions: they support the dynamics of survival. The new brain, or neocortex, is highly developed only among human beings. It is therefore recent in evolutionary terms. It develops late at an individual level and does not reach maturity before adulthood. It is made of trillions of interconnected neurons. It is often called the associative brain and can be presented as a supercomputer.

In different terms, one can contrast higher brain functions and lower brain functions. It is also commonplace to refer to the emotional-instinctive brain and to the brain of the intellect, knowing that this distinction is simplistic since emotional states influence neocortical activity and neocortical activity can influence emotional states.

Anyway the human brain – the most complex structure in the universe that we know about – cannot be described in a few words. In practice, the important point is to think in terms of chronology. We must first wonder at what phase of brain development birth takes place. It is essential to realize that the birth of a human baby occurs at an important phase of development of the primitive brain and early with regards to the development of the neocortex. Thinking in terms of chronology leads us to recall that the human left and right cerebral hemispheres are anatomically and functionally asymmetric and that the development of the right hemisphere precedes the development of the left one.[2,3] During the perinatal period, the right hemisphere is already in a phase of well-advanced fast development, but the left hemisphere shows a growth spurt starting in the second year after birth. The comparatively fast development of the right hemisphere is all the more important since there is today an accumulation of data suggesting that the right hemisphere is more involved than the left hemisphere in emotional processing.[4]

At a time when our cultural lack of understanding of the basic needs of labouring women has reached an extreme degree, which is also the age of synthetic oxytocin, we have reasons to anticipate a tendency towards a disharmonious development of the human brain. When considering the future of *Homo sapiens*, we must also take into account the fact that there are now widely used substitutes for the oxytocin system to give birth and to feed babies, and that this system has strong links with all other basic physiological systems. We must raise questions about the trans-generational effects of acquired traits, such as those related to underused organs or physiological functions. We must also keep in mind that, in the age of simplified fast techniques of caesarean section, the tendency towards an increased brain volume – in practice an increased volume of the neocortex – can be transmitted to the following generations.

After considering all the factors that made the perinatal period the most highly disturbed phase of human life, we can wonder what kind of evolutionary tendencies we might expect.

Towards the planet of Aspies?

Since the transition between intra-uterine and extra-uterine life occurs at a critical phase of development of the emotional brain and precedes the growth spurt of the brain of the intellect, it is plausible that the average ratio of 'emotional quotient' (EQ) to 'intellectual quotient' (IQ) will change in the future. One can expect the advent of different dominant personality traits. A gradual reduction in EQ seems probable. This would mean a reduced ability to reach and to express certain emotional states, a reduced ability to understand the emotions of others, a lack of empathy, a lack of nonverbal communication skills, some particularities in terms of neuromotor profile and different basic states of health, since one cannot dissociate primitive brain structures, the endocrine system and the immune system. Conversely, it seems possible that the development of the brain of the intellect is not routinely hindered in spite of defective 'foundations'. It might even reach an increased degree of development related to an increased volume.

When considering this composite drawing of *Homo sapiens* of the future, one cannot help thinking of the 'Aspies'. This is a term used by people identifying with Asperger's syndrome, when referring to themselves in casual conversation. It seems that their number is increasing, to such a degree that a real subculture of aspies has formed, thanks in particular to websites like Wrong Planet. It is significant that Wrong Planet, although limited to the English language, had more than 70,000 members at the end of 2012. Such websites confirm the contrast between the inability to pick up the body language of other people and a huge networking capacity, as long as written language is used. When reading texts from websites and books written by aspies, it is clear that many aspies are talented writers. One of the main characteristics of texts found in the Community Forum of Wrong Planet is their grammatical perfection. The participants have obviously reached a degree of cognitive and linguistic development far above average.

Today Asperger's syndrome is considered an 'autistic spectrum disorder'. This implies that all autistic traits are usually interpreted in the framework of pathological conditions, rather than as personality traits. There is therefore a tendency to focus on deficiencies. However, the selective advantages can be enormous. Aspies can have intense preoccupation with a narrow subject and an unusual capacity for deep concentration. Aspies are attracted less to people and emotions but more to factual patterns. According to Simon Baron-Cohen, director of the Autism Research Centre at the University of Cambridge, people with autism naturally think like scientists, and there is a link between autism and mathematical talent. An American study, reported in several prestigious scientific journals, examined 11,000 students across the USA and found that more young adults with an autistic spectrum disorder choose scientific topics than their peers in the general population.[5] It is not surprising that Asperger's syndrome and aspie lifestyle are comparatively common among top, highly-specialized scientists. Vernon Smith, an aspie who received the Nobel Prize in 2002 for inventing the field of experimental economics, wrote about himself: 'I can switch out and go into a concentrated mode and the world is completely shut out'.

Since Asperger's syndrome is considered an autistic spectrum disorder, we must emphasize that our composite drawing of *Homo sapiens* of the future is compatible with epidemiological studies looking at the genesis of autism. In spite of a great diversity of research protocols in a great diversity of countries involved, the most significant studies of autism included in the Primal Health Research Database detect risk factors in the perinatal period. This is the case in a study of the entire Swedish population born over a period of twenty years,[6] in an Australian study comparing 465 cases with siblings and controls,[7] in a Japanese study evaluating the risk of autism in relation to the hospital where the subjects were born,[8] and in an Israeli study based on the concept of perinatal suboptimality scores.[9] An exploration of the database via the keyword autism suggests that, for this particular disorder, the period of transition between intra-uterine and extra-uterine life is critical for gene-environment interaction.

Studies of the oxytocin system of autistic subjects suggest interpretations of the perinatal period as a critical period in the genesis of the autistic spectrum disorder. The first clues came from a study of midday blood samples from twenty-nine autistic and thirty age-matched normal children.[10] The autistic group had significantly lower blood oxytocin levels than the normal group. Oxytocin increased with age in the normal but not the autistic children. These results inspired an in-depth inquiry into the oxytocin system of autistic children. In recent years it has become clear that oxytocin can appear in the brain in several forms. There is the nonapeptide oxytocin (OT) and the 'C-terminal extended peptides', which are described together as OT-X. The OT-X represent intermediates of oxytocin synthesis that accumulate due to incomplete processing. Twenty-eight male children diagnosed with autistic disorder were compared with thirty-one age-matched non-psychiatric control children: there was a decrease in blood OT, an increase in OT-X and an increase in the ratio of OT-X/OT in the autistic sample, compared with control subjects.[11] In other words, autistic children show deficits in the processing of oxytocin. These findings have justified studies of the possible therapeutic effects of oxytocin in autism. This has been possible since it is now well known that oxytocin given through nasal sprays can reach the brain receptors.

If there is a future for autistic traits, we must also consider the documented association of such personality traits with conditions other than those of the autistic spectrum disorder. This is the case with anorexia nervosa.

Several teams of psychiatrists, such as the team of Janet Treasure of the Institute of Psychiatry of King's College Hospital, in London, have emphasized the importance of autistic traits in anorexia nervosa.[12,13] People with anorexia nervosa find it difficult to change self-set rules; they see the world in close-up detail, as if they were looking through a zoom lens, and risk getting constantly lost in the details. Christopher Gillberg, and the team of the Department of Child and Adolescent Psychiatry at Goteborg University in Sweden, found that 23% of female patients with severe eating disorders had symptoms of the autistic spectrum.[14] I became aware of these clinical observations after studying both autism and anorexia nervosa from a primal health research perspective. Their main risk factors are similar in terms of timing.[15,16] I was testing the hypothesis that when two pathological conditions or personality traits share the same critical period for gene-environment interaction, we should expect further similarities, particularly from clinical and pathophysiological perspectives.[17] Starting from these considerations, I found that the two conditions did indeed have other similarities.

There are probable similarities regarding the oxytocin system. It has been reported that the level of oxytocin in the cerebrospinal fluid of women with 'restricting anorexia' is significantly lower than the level of oxytocin in bulimic and control subjects.[18]

Similarities between autism and anorexia nervosa are also revealed by explorations of brain function, particularly brain scanning and electroencephalography. There are concordant data suggesting that in autistic disorders[19,20] and in anorexia,[21,22,23,24] there is a left hemisphere preponderance (or right hemisphere deficit). Mirror neuron dysfunction has been demonstrated in autism spectrum dysfunction.[25,26] Studies of the mirror neuron system in anorexia would be welcome.

After studying anorexia nervosa and autism in parallel, one can echo the suggestion, expressed by some psychiatrists, that anorexia nervosa might be considered a female variant of the autistic spectrum. A plausible interpretation of why it is undoubtedly more female is that antenatal exposure to male hormones might protect against the expression of anorexia nervosa. Such an interpretation is suggested by a study of twins.[27] Girls who had a twin brother were at low risk of anorexia nervosa compared with girls who had a twin sister and with controls. This interpretation is reinforced by the negative results of genetic linkage analyses that could not detect any change on the X chromosome.[28]

It is difficult to draw the borders of the world of aspies.

The future of depression

Huge studies among American college students (in post-secondary institutions) provide food for thought about the current transformations of *Homo sapiens*. We have already mentioned the synthesis of seventy-two studies according to which college graduates in 2009 were 40% less empathetic than those in 1979.[29] The decline has been progressive, and particularly fast after the year 2000. We compared the results of this study with birth statistics, breastfeeding statistics and data about genital sexuality to suggest that dysfunction of the oxytocin system, which has strong links to other vital physiological functions, is responsible.

Depression is another hot topic among college students. A group of American post-secondary institutions (148 in 2011) participate in the National Health Assessment of students (116,254 surveys were completed in 2011). When comparing the data from 2000 (http://www.acha-ncha.org/docs/ACHA-NCHA_Reference_Group_Report_Fall2000.pdf) and the data from 2011(http://www.acha-ncha.org/docs/ACHA-NCHA-II_ReferenceGroup_DataReport_Spring2011.pdf), it appears that the rate of students reporting that they have been diagnosed with depression has increased from 10% to 21% in eleven years! Genetic factors and changes in diagnostic criteria used by doctors cannot explain these differences. In order to interpret such data, we must first think about changes in lifestyle that have occurred and should be considered when comparing students born in around 1980 and students born in around 1991. There is one inescapable answer. During that decade the history of childbirth sped up. There were new steps in the evolution of the deep-rooted cultural lack of understanding of the physiological processes of birth. In other words, it was a time when the number of women who were able to give birth to their baby and to the placenta thanks only to the release of their natural hormones dramatically decreased.

The biology of depression is complex and still partly mysterious. The genesis of 'clinical depression' seems to be related to the way the physiological set point levels are established during the primal period. For example, depressed people have much higher levels of cortisol than others during the recovery period after stressful life events.[30]

This seems to be related to the way the set point levels of the hypothalamo-pituitary-adrenal axis have been adjusted at the beginning of life. There is not space here to discuss the dominant theories of depression. For now simply consider that all the brain areas that show altered activity in depressed subjects are in a phase of development/adjustment during the period surrounding birth. This is the case of the raphe nuclei (the source of serotonin), the suprachiasmatic nucleus (which controls the sleep/wake cycle), the ventral tegmenta area (a critical part of the brain's reward system), the nucleus accumbens (which regulates dopamine release), the anterior cingulate cortex (cases of severe depression have been successfully treated by inactivation of this brain zone) and the subgenual cingulate (a region rich in serotonin transporters). It makes sense that the development of these primitive brain structures is influenced by perinatal events.

There is a lack of scientific studies confirming that the period of transition between intra-uterine and extra-uterine life might be critical for gene-environment interaction relevant to the origin of clinical depression. There is a tendency to focus on alterations of brain development associated with 'less than optimal early maternal care', without evaluating the effects of perinatal events on the quality of early maternal care.[31] It is certain that 'less than optimal early maternal care' is independently a risk factor for depression. This was demonstrated in particular by a Finnish study of people who had been separated at birth, until an average age of seven months, from parents who had tuberculosis. This population was at an increased risk for hospital-treated depression in adulthood.[32] Developmental neurosciences often focus on early preverbal development of emotion processing, and also on the later development of verbal cognitive processes. The importance of the perinatal cataclysm with its multiple variants is probably underestimated.

Even if the physiological mechanisms are still ill-defined, we now have a sufficient accumulation of data at our disposal to regard depression as a probable leading cause of disability in the future.

The sorcerer's apprentice

By interfering routinely, at a planetary level, with the development of vital physiological functions, *Homo sapiens* is playing the role of the sorcerer's apprentice. As long as the sorcerer's apprentice is in action, one can only expect the unexpected.[33]

This is why looking at the future of *Homo sapiens* is frightening. The sorcerer's apprentice has lost the concept of the interconnectedness of all things. Current classifications of personality traits and mental disorders might become useless because they are too simplistic. When considering disharmonious developments of primitive brain structures versus new brain structures, and of right hemisphere versus left hemisphere, we were led to consider the increased number of aspies on the one hand and of depressed people on the other hand. We must keep in mind, however, that the reality is complex. A person can be diagnosed with both Asperger's syndrome and depression, or with both Asperger's and Attention Deficit Hyperactivity Disorder (ADHD), or with both Asperger's and Obsessive Compulsive Disorder (OCD).

The framework of disharmonious brain development is intricate.

We must also keep in mind that the primitive brain is not an isolated organ. Apart from its interconnectedness with the new brain, it cannot be dissociated from our basic adaptive systems. Today one cannot separate nervous system, endocrine system and immune system. In practice this means that the concept of personality traits should include the concept of metabolic types. We must break down the frontiers of conventional medical specialization to unveil links between pathological conditions. For example, it is becoming difficult to separate schizophrenia and type-2 diabetes. People with diabetes have an increased incidence of mood disorders and schizophrenic people are at increased risk of type-2 diabetes. It has been demonstrated that insulin also regulates the brain's supply of dopamine, a neurotransmitter with roles in attention, reward, motivation and motor activity. The relationship between ADHD and obesity offers an opportunity to illustrate the need to mix the concepts of personality trait and metabolic type: the average weight of people with ADHD is comparatively high.[34]

We have learned from ancient wisdom how the actions of the sorcerer's apprentice can be neutralized and even reversed. This should lead us to rediscover the laws of nature: this should lead us to rely on the *physiological* perspective.

CHAPTER EIGHTEEN

The future of enthusiasm

If 'Enthusiasm moves the world', as the philosopher-politician Arthur Balfour has asserted, the future of *Homo sapiens* and the future of enthusiasm are parts of the same question.

Enthusiasm is not included by scientists in conventional lists of basic emotional states and emotional traits. As suggested by the roots of the word, an enthusiastic person is inspired by divinities. Socrates taught that the inspiration of poets is a form of enthusiasm. Enthusiasm is a way to transcend the objective daily reality. It can be presented as a variant of transcendent emotional states, the ultimate form of which gives access to an out of space and time reality. The associative neocortical brain can only give access to space and time reality. Therefore, when we have access to another reality it is through emotional states. Since the role of religions is to channel the universal transcendent emotional states, it is no wonder there have always been strong links between the concept of enthusiasm and the concept of religion. For example, 'The Enthusiasts' was the name of a Syrian religious sect of the fourth century; Protestant sects of the sixteenth and seventeeth century were also called 'enthusiastic'.

Keeping in mind that the routinely disturbed transition between intra-uterine and extra-uterine life occurs during a crucial period of the development of the emotional brain, we cannot help raising questions about the future, at a cultural level, of transcendent emotional states in general. These questions go far beyond the future of religious institutions. They are first and foremost about the origin of inspiration, and therefore scientific and artistic creativity. Like enthusiasm and transcendent emotional states in general, inspiration is not frequently studied by modern psychology. The mysterious burst of creativity called inspiration has always been considered a divine spark. In Greece the Muses were the goddesses of inspiration in literature, science and the arts.

What is the future of transcendental emotional states? What is the future of a hyper-brainy *Homo sapiens* deprived of enthusiasm and inspiration?

CHAPTER NINETEEN

Homo sapiens and the virosphere

Let us combine two issues. One is the anticipated emotional dysregulation of the genus *Homo*. The other is the relationship of humanity to the world of viruses. Linking these issues would imply that we have overcome the unique kind of blindness generated by overspecialization.

Smashing barriers

When considering the possible effects of routine risky interference in the development of primitive brain structures, we emphasized that the conventional barriers between the nervous, endocrine and immune systems get in the way. We reintroduced the term 'primal adaptive system'.[1] Suggesting that the concept of personality traits should include the concept of metabolic types was a step towards understanding the unity of the basic adaptive systems involved in what we commonly call health. Another step is to claim that the quality of our immune system should not be dissociated from other personality traits.

Today we must smash the barriers between the functions of the primitive brain structures and the functions of the immune system. For example, the activities of the immune system are dependent on the release of cortisol by the adrenal gland: it is well known that cortisol levels are under the control of the hypothalamus, a part of the primitive brain that works as an endocrine gland. The 'set point levels' of the hypothalamus are established during critical phases of the primal period. We now know that stimulating the immune system sends a flow of information to the hypothalamus. Some antigens – substances that stimulate the immune system – can considerably increase the electrical activity of certain nerve cells of the hypothalamus. So the immune system can be seen as an actual sensory organ which gives information to the brain.

Some spectacular experiments on the conditioning of immune reactions have demonstrated that a signal via the nervous system can affect immune functions. They confirm the marriage of the nervous system and the immune system as a theoretical breakthrough. Robert Ader and Nicholas Cohen gave rats saccharinated water at the same time as a drug (Cytoxan) that depresses the immune system and triggers digestive troubles. After that they could depress the immune system of these rats, and even kill them, by giving them saccharinated water alone. Further experiments of Pavlovian conditioning of the immune system confirmed these results.[2]

Since modern science is able to point to the unity of the primal adaptive system, we must raise questions about the effects of a disharmonious development of the human brain on immunological functions in the same way that we have raised questions about the possible effects on emotional regulation.

In order to anticipate such effects we must first recall that the primary function of the immune system is to detect a wide variety of agents, from viruses and bacteria to parasitic worms, and to distinguish them from the organism's own healthy tissues. If this primary function is altered the immune system may attack normal tissues as if they were foreign organisms. This is the mechanism of autoimmunity, leading to diseases such as type-1 diabetes, rheumatoid arthritis, Hashimoto's hypothyroidism and lupus erythematous. The prevalence of autoimmune diseases such as type-1 diabetes is increasing[3] and a caesarean birth appears to be a risk factor.[4,5] A dysregulated immune system can also overreact: the effects of such overreactions are allergic diseases. The prevalence of allergic diseases including asthma is increasing, and antibiotics in the perinatal period and caesarean section appear to be risk factors.[6,7,8,9,10] The results of a study about the risks of asthma in childhood in relation to exposure to antibiotics at the end of intra-uterine life suggest the probable intermediate role of the gut flora.[11]

Immune function can also be depressed and lack power to neutralize a great variety of pathogenic agents and cancer cells. It is impossible to list all possible groups of pathogenic agents here; thus this discussion is confined to explaining the reasons for focusing on emerging virus diseases as a major threat to human health.

The viral threat

A small number of established facts should suffice to convince anyone that the survival of our species is highly dependent on our relationship with the world of viruses.

Viruses represent the largest proportion of biomass if one takes into account an almost infinite number of viruses in the oceans. The main threat comes from viruses with an RNA genome. Such viruses are pieces of RNA surrounded by proteins and often enveloped in a membrane. They can only reproduce inside infected cells. Since they are capable of extremely rapid mutation they have an enormous capacity to adapt immediately to their environment and particularly to changes in climatic conditions. Compared with other living organisms, their capacity to adapt is of a different order of magnitude. The emergence of hantavirus pulmonary syndrome following an El Niño event that started in summer 2009 is an illustration of the capacity of viruses with RNA genomes to adapt quickly to differences of temperature. The 'El Niño-Southern Oscillation' is a quasi-periodic climate pattern that occurs across the tropical Pacific Ocean roughly every five years. It causes variations of temperature at sea level. Proper investigation of such events is a vital consideration at a time when significant climatic changes are expected during the next few centuries, whatever the comparative role of solar variability and human activities.

When evaluating the viral threat we must take into account that frequent air travel is a new phenomenon. It is undoubtedly a major contributing factor in the spread of emerging diseases. This is how one can explain the rapid spread of the severe acute respiratory syndrome (SARS) coronavirus in 2003, when an outbreak in Hong Kong nearly became a worldwide epidemic, with more than 8,000 cases and 774 deaths worldwide, according to the World Health Organisation. Within weeks, SARS (an RNA virus) spread from Hong Kong to infect individuals in thirty-seven countries. In fact, the capacity of highly pathogenic viruses with an RNA genome to spread out all over the world has preceded the age of frequent air travel. According to recent estimates, the 1918 world flu epidemic killed as many as 100 million people, or 5% of the world population at that time.[12]

Analysing the viral threat is an elegant way of realizing the limits of our domination of nature, as a basic strategy for survival that started with the Neolithic revolution. This domination has included interferences in human physiological processes, particularly those related to sexuality, and especially the birth process. An emotionally dysregulated human being is a product of such strategies for survival. This kind of human being is unable to shift from knowledge to awareness. He is blind to the effects of his behaviour and continues to dramatically disturb ecosystems without taking into account that viruses can adapt much more easily and quickly than mammals to drastic environmental changes, particularly climatic changes. What kind of *Homo* can break this vicious circle?

Let us repeat that the sorcerer's apprentice must be ready to expect the unexpected. Although the viral threat should be prioritized ahead of many other considerations, because the evaluation of its importance is based on indisputable pieces of knowledge, one cannot ignore concomitant threats.

We must also keep in mind that there are millions of species of fungi in the world, but only 100,000 have been identified. Reports of new types of fungal infection in plants and animals have risen nearly tenfold since 1995. It has been suggested that climate change is a culprit. This threat might be mostly indirect, through the destruction of staple crops. The Irish potato famine of the 1840s showed how devastating such pathogens can be. In fact, the risks must be put into perspective, because fungi are not spread as easily from person-to-person as viruses, and anti-fungal agents can effectively tackle most infections. As Matthew Fisher, an emerging-disease researcher at Imperial College London, said: 'I'd be very surprised if an abrupt fungal infection killed a large swathe of people. But it's not impossible.'[13]

We are now entering a new phase in history. From now on, the main topic should not be the history of the relationships between humans, but the history of the relationship between humanity and Mother Earth. The virosphere is a major component of Gaia… of Mother Earth as an organism with its own self-regulatory functions.[14]

CHAPTER TWENTY

Cultural blindness

Imagine that you are sitting in a conference room during a public lecture. The topic is childbirth. You wonder who the other participants are. It is clear that there are many pregnant women and midwives, and perhaps also some doctors. Obviously participants have a direct immediate interest in childbirth, either for professional or for personal reasons. From the interaction between the speaker and the public, it appears that this event is dominated by short-term considerations. It is not perceived as a discussion about the future of humanity.

There is a universal lack of interest in the long-term consequences of how babies are born. In the current scientific context, this can be presented as kind of blindness. It is as if there are cultural forces that are pushing us to ignore the essential.

Cul-de-sac epidemiology

One of the many facets of this cultural blindness is what we have called 'cul-de-sac epidemiology'.[1] This encompasses research about topical issues. Despite the publication of this research in authoritative medical or scientific journals, the findings are shunned by the medical community and the media.

In 2000 we drew attention to a Swedish study, published in 1990 by Bertil Jacobson, leading to the conclusion that certain obstetric drugs are risk factors for drug addiction in adult offspring.[2] The results have never been confirmed or invalidated by other teams of researchers. Yet drug addiction is one of the main preoccupations of our time. We took as other examples studies of risk factors for autism in the period surrounding birth. At the end of his life, the Nobel Prize winner Niko Tinbergen studied autistic children with the methods of a field ethologist. He came to the conclusion that there are risk factors for autism in the perinatal period, such as anaesthesia during labour and induction of labour.[3] His observations inspired a study by Ryoko Hattori (Kumamoto, Japan) published in 1991.[4] She found that the 'Kitasato University's method' of delivery is a risk factor for autism. This method is characterized by a combination of sedative and analgesic agents, together with a planned delivery induced a week before the due date. Since that time, there have been several valuable large studies published in authoritative medical journals.[5,6,7] All of them detected risk factors for autism in the perinatal period.

Furthermore, these studies also reported negative findings about those events preceding birth which *do not* significantly influence the risks (same average birth weights, placental weights and head circumferences at birth, and no influence on the risk of autism of maternal disorders such as pre-eclampsia). The negative findings also include what happens after the birth, such as the mode of infant feeding and nature of vaccinations. There is no example, in particular, of a valuable epidemiological study detecting correlations between MMR (Measles, Mumps, Rubella) vaccination and the diagnosis of autism, or one detecting correlations with a vaccine containing a mercurial derivative.[8,9,10,11,12] However, thousands of articles all over the world have discussed the theory that certain vaccinations in infancy are risk factors for autism. At the same time there has been a lack of interest in huge valuable studies published by reliable epidemiologists about risk factors in the perinatal period.

This lack of interest – this blindness – is cultural. It is shared by the media, the general public and even authoritative medical and scientific circles. This is well illustrated in the following example. A review article about autism – including the epidemiology of autism – was published in a prestigious medical journal.[13] This article was followed by 120 references. The studies detecting risk factors at birth were not mentioned. This shows the degree to which the 'blindness' has pervaded medical circles. As an example of this same blindness in scientific circles, consider a long, detailed article published in the prestigious journal *Nature*.[14] The objective was to explain how researchers are 'digging into the myriad causes of autism' to refine its definition and find elusive biological signatures. Again the studies detecting risk factors at birth were not mentioned.

Cul-de-sac epidemiology is responsible for our current difficulty in offering plausible interpretations of recent phenomena. Once more we can take as an example the increasing prevalence of autism, something which appears to be happening faster in some countries. Bai Xueguang, a professor of neurology at the People's Hospital of Hubei Province, based in Wuhan City, has revealed that the number of children with autism was growing at an annual rate of 20% in China, much more quickly than the world average. (http://news.xinhuanet.com/english/2004-08/11/content_1759576.htm) Why in China? Those who are familiar with the Primal Health Research Database and who are also aware of the evolution of Chinese birth statistics will immediately consider the data provided by Bai Xueguang as a way to support the results of epidemiological studies.

Epidemiology has been presented as a typical example of a scientific discipline whose potential is still limited by a cultural lack of interest in the way babies are born. We might have chosen the example of demography, the statistical study of the evolution of populations. Population scientists look towards the future through extrapolations from available data. They take into account factors that can influence the evolution of a population, such as the degree of fertility. I have never heard of demographers raising questions about, for example, the fertility of a human population born by caesarean section.

Learning from biographers

Another way to evaluate the extent of this cultural blindness is to look at the biographies of well-known influential pioneers in human sciences or medicine. Some of them have been interested in the way babies are born, and in many cases this was the part of their work that was ignored by their contemporaries and thus has not been transmitted to following generations. The work of Maria Montessori is a typical example. It is possible to read the most detailed of her numerous biographies in a great variety of languages without even a hint of the importance she attached to the way babies are born. This is the beginning of the chapter 'The newborn child' in her book entitled *The Secret of Childhood*,[15] (which is an adaptation in the English language of a book originally published in French in 1930):

> At birth a child does not enter into a natural environment but into one that has already been extensively modified by men. It is an alien environment that has been built up at nature's expense by men... At no other period of his life does a man experience such a violent conflict and struggle, and consequent suffering, as at the time of birth. This is a period that certainly deserves to be seriously studied, but as yet no such study has been made...

We might make similar comments about biographies of Raymond Dextreit, who after the Second World War was an influential proponent of 'Natural Medicine'. Writing in the French journal *Bionaturism* in 1954, Dextreit commented on birth:

All the cares lavished on the baby from the moment it appears only serve to unbalance it at its point of departure. When the baby arrives there is a shock waiting for it. More, a series of shocks. After several months in the calm, silent, dark and comfortable softness of the mother's womb, the baby is suddenly and abruptly made aware of noise and light in an ambient temperature much lower than that it has known during those nine months. As if this rush of shocks was not enough, it is then exposed to the particularly brutal treatment of early cutting of the umbilical cord... Instead of thinking about vital precautions, we concentrate on microbes. Instead of trying to protect and develop all the marvelous natural immunities, we kill all the microbes... For the moment we will go no further, hoping perhaps presumptuously that we have drawn the reader's attention to one aspect of the human problem which men of science believe they have solved.

Dextreit was right to introduce the word presumptuously. Just as individual pioneers have had significant parts of their work ignored, so have scientific or medical teams had their work relating to childbirth overlooked. The Vienna Psychoanalytic Society offers a typical example. Freud considered Otto Rank, who worked closely with him for twenty years, the most brilliant of his Viennese disciples. The turning point was in 1924, when Rank dared to publish *Das Trauma der Geburt* (The Trauma of Birth),[16] exploring how art, myth, religion, philosophy and therapy were illuminated by the period of transition between intra-uterine and extra-uterine life. Otto Rank had put his finger on the foundation of our civilization.

The vital function of madness

Since the blindness regarding scientific investigation of childbirth is cultural, we must rely on counter-cultural ways of thinking and being to herald a new era with renewed strategies for survival. This is undoubtedly difficult, since human beings are endowed with gregarious tendencies. Like most other primates we need to live in groups. Thanks to our capacity to communicate in sophisticated ways, particularly through language, we create cultural milieus. Cultural milieus prescribe and proscribe behaviour; they dictate what people should and should not do given their social surroundings and circumstances. Norms are established, with at least some degree of consensus, that are enforced through social sanctions.

Because antisocial behaviours are contrary to the standards of our societies, they are easily considered pathological or morally unacceptable. In such a context it is difficult to accept that we might take advantage of certain deviations. Among these deviations we must mention eccentrics and geniuses. Most highly creative, legendary geniuses had unusual, eccentric personalities and manifested many schizotypal traits. Isaac Newton lived most of his life alone. Albert Einstein had poor grooming and hygiene and well-documented interpersonal deficiencies. Bertrand Russell was an aloof, lonely and somewhat insecure child, and led an unstable adult life.

The link between 'genius' and 'insanity' has been widely studied since the nineteenth century, after Cesare Lumbroso, an Italian psychiatrist, published his book *L'uomo di genio* in 1864, and Francis Galto published *Hereditary Genius* in 1869. The personality traits of geniuses might simply be written off as unimportant details, but for the fact that their family history indicates strong connections between genius and schizophrenia. Many studies, such as one carried out in Iceland, have confirmed that the relatives of creative people suffer an increased rate of schizophrenia.[17] It is noticeable that James Joyce's daughter was schizophrenic, and the family pedigree of Bertrand Russell was loaded with schizophrenic people: his uncle William was 'insane'; his aunt Agatha was delusional; his son John was diagnosed with schizophrenia and his granddaughter Helen also suffered from schizophrenia and committed suicide by setting fire to herself. Albert Einstein's son by his first marriage suffered from schizophrenia. The son of John Nash, the gifted schizophrenic mathematician and Nobel Laureate in Economics, suffers from schizophrenia too. A few schizophrenic individuals, but many first-, second- or third-degree relatives, who share part of the schizophrenic genome, are some of the most creative individuals around. The close relationship between madness and creativity led David Horrobin to assume that 'schizophrenia shaped humanity'.[18]

Some personality traits – at the limit of what is considered pathological – are associated with the capacity to overcome inhibitions acquired during the process of cultural integration. Such traits might be reinforced after a certain age. If we take a positive and realistic approach, we can claim that there are people who can overcome our cultural blindness more easily than others. These people are more precious than ever. We must listen to them. I am thinking of Wilhelm Reich, who was originally a disciple of Freud. At the end of his life he was considered insane according to conventional criteria. His behaviour was not compatible with cultural norms. It is significant that he died in jail. It is also significant that his last book was *The Murder of Christ*. In the vocabulary of Reich 'Christ' is the realization of Natural Law. The book is mostly about the domination, control, and repression of human physiological processes related to sexuality. Its counter-cultural conclusion is that we have to 'turn the tide toward concentration on the essential of human life' and that 'the concern for the welfare of the newborn baby... can be surpassed by nothing on earth'.[19]

I am also thinking of Frédérick Leboyer, whose way of thinking and lifestyle are – to say the least – unconventional. In his famous book *Birth Without Violence*, originally published in 1974, he wrote:

...One should have to be naïve indeed to believe that so great a cataclysm would not leave its mark. Its traces are everywhere;

on the skin, in the bones, in the stomach, in the back, in all our human folly, in our madness, our tortures, our prisons, in legends, epics and myths...

The most common way to distort the misunderstood message of Leboyer has been to associate his name with the word 'method'.[20]

What about the future? Can we hope that there will be a sufficient number of 'crazy' influential people to overcome the current cultural blindness? Let us dream of an imaginary scenario. Let us imagine, for example, that the eccentric traits of the Health Minister of a huge 'developed' country have become more pronounced with age. After considering birth statistics, she notices a tendency towards more forceps or ventouse deliveries, a more widespread use of pharmacological assistance, an increased rate of caesareans, without any improvement regarding the number of babies alive and healthy at birth.

She is convinced that the priority is to create situations compatible with easy birth. She is inspired by her personal experience as a mother and also by the way doulas were selected for a randomized controlled trial in Houston, Texas.[21] This is how she has designed a radical plan to be approved by a committee. The basis of her plan is that to become an obstetrician or a midwife the prerequisite will be to be a mother who has a personal positive experience of unmedicated birth. How will the committee react to such a project? The most probable scenario is that the Minister of Health will be considered insane and urgently replaced by a colleague capable of coping with important matters.

However, if the time is ripe for an emerging new awareness, another scenario should not be dismissed.

Let us imagine that most committee members are able to listen to the bizarre and eloquent Minister of Health and to overcome their blindness. They become convinced that, where childbirth is concerned, 'we are like a traveller finding out that she or he is on a wrong path or in a cul-de-sac'. In this case the best action is to go back to the point of departure and to restart the journey in another direction.

The conclusion will be: let us get rid of the aftermath of thousands of years of beliefs and rituals and restart from the physiological perspective. Let us act as if it is not too late.

REFERENCES

Introduction

1. Odent, M. *Primal Health*, Century Hutchinson, London, 1986.
2. Li, H-T., Ye, R., Achenbach, T., Ren, A., Pei, L., Zheng, X., Liu, J-M. 'Caesarean delivery on maternal request and childhood psychopathology: a retrospective cohort study in China', *BJOG* 2010; DOI: 10.1111/j.1471-0528.2010.02762.x.
3. Hultman, C., Sparen, P., Cnattingius, S. 'Perinatal risk factors for infantile autism', *Epidemiology* 2002; 13: 417-23.
4. Glemma, E.J., Bower, C., Petterson, B., et al. 'Perinatal factors and the development of autism', *Arch Gen Psychiatry* 2004; 61: 618-27.
5. Hattori, R., et al. 'Autistic and developmental disorders after general anaesthetic delivery', *Lancet* 1991; 337: 1357-1358 (letter)
6. Stein, D., Weizman, A., Ring, A., Barak, Y. 'Obstetric complications in individuals diagnosed with autism and in healthy controls.' *Compr Psychiatry* 2006 Jan-Feb; 47(1):69-75.
7. Cnattingius, S., Hultman, C.M., Dahl, M., Sparen, P. 'Very preterm birth, birth trauma and the risk of anorexia nervosa among girls'. *Arch Gen Psychiatry* 1999 56: 634-38.
8. Favaro, A., Tenconi, E., Santonastaso, P. 'Perinatal factors and the risk of developing anorexia nervosa and bulimia nervosa'. *Arch Gen Psychiatry* 2006 63(1):82-8.
9. Pistiner, M., Gold, D.R., et al. 'Birth by cesarean section, allergic rhinitis, and allergic sensitization among children with a parental history of atopy'. *J Allergy Clin Immunol.* 2008 Aug;122(2):274-9. Epub 2008 Jun 20
10. Bager, P., Wohlfahrt, J., Westergaard, T. 'Caesarean delivery and risk of atopy and allergic disease: meta-analyses'. *Clin Exp Allergy*, 2008. Apr;38(4):634-42.
11. Xu, B., Pekkanen, J., Hartikainen, A.L., Järvelin, M.R. 'Caesarean section and risk of asthma and allergy in adulthood'. *J Allergy Clin Immunol.* 2001 Apr;107(4):732-3.
12. Roduit, C., Scholtens, S., de Jongste, J.C., et al. 'Asthma at 8 years of age in children born by caesarean section'. *Thorax* 2009 Feb;64(2):107-13. doi: 10.1136/thx.2008.100875. Epub 2008 Dec 3.
13. van Nimwegen, F.A., Penders, J., Stobberingh, E.E., et al. 'Mode and place of delivery, gastrointestinal microbiota, and their influence on asthma and atopy'. *J Allergy Clin Immunol.* 2011 Nov;128(5):948-55.e1-3. Epub 2011 Aug 27.
14. Huh, S.Y., Rifas-Shiman, S.L., Zera, C.A., et al. 'Delivery by

caesarean section and risk of obesity in preschool age children: a prospective cohort study'. *Arch Dis Child.* 2012 Jul;97(7):610-6. Epub 2012 May 23.

15. Turnbaugh, P.J., Ley, R.E., Mahowald, M.A., et al. 'An obesity-associated gut microbiome with increased capacity for energy harvest'. *Nature* 21 December, 2005; 444:1027-1031.

16. McKinney, P.A., Parslow, R., Gurney, K.A., Law, G.R., Bodansky, H.J., Williams, R. 'Perinatal and neonatal determinants of childhood type 1 diabetes. A case-control study in Yorkshire, U.K.' *Diabetes Care* 1999 Jun;22(6):928-32.

17. Cardwell, C.R., Stene, L.C., Joner, G., et al. 'Caesarean section is associated with an increased risk of childhood-onset type 1 diabetes mellitus: a meta-analysis of observational studies'. *Diabetologia* 2008 May;51(5):726-35. Epub 2008 Feb 22.

18. Raine, A., Brennan, P., Medink, S.A. 'Birth complications combined with early maternal rejection at age 1 year predispose to violent crime at 18 years'. *Arch. Gen. Psychiatry* 1994; 51: 984-88.

19. Salk, L., Lipsitt, L.P., et al. 'Relationship of maternal and perinatal conditions to eventual adolescent suicide'. *Lancet* March 16th, 1985, pp 624-27

20. Jacobson, B., Nyberg, K., et al. 'Perinatal origin of adult self-destructive behavior'. *Acta. Psychiatr. Scand*, 1987; 76: 364-71

21. Jacobson, B., Bygdeman, M. 'Obstetric care and proneness of offspring to suicide as adults: case control study'. *BMJ* 1998; 317: 1346-9

22. Jacobson, B., Nyberg, K. 'Opiate addiction in adult offspring through possible imprinting after obstetric treatment'. *BMJ* 1990; 301: 1067-70

23. Nyberg, K., Buka, S.L., Lipsitt, L.P. 'Perinatal medication as a potential risk factor for adult drug abuse in a North American cohort'. *Epidemiology* 2000; 11(6): 715-16.

Chapter 1: *Ecce Homo*

1. Cunnane, S. 'Iodine: The Primary Brain Selective Nutrient' in: Stephen Cunnane, *Survival of the fattest: the key to human brain evolution*, World Scientific Publishing, Singapore, 2005.

2. Stagnaro-Green, A., Sullivan, S. 'Iodine supplementation during pregnancy and lactation', *JAMA* 2012;308(23): 2463-64.

3. Crawford, M.A., Marsh, D. *The driving force: Food in Evolution and the Future*, William Heinemann, London, 1989.

4. Morgan, E. *The Aquatic Ape*, Souvenir Press, London, 1982.

5. Morgan, E. *The Scars of Evolution*, Souvenir Press, London, 1990.

6. Morgan, E. *The Descent of the Child*, Souvenir Press, London, 1994.

7. Vaneechoutte, M. (Ed.) *Was Man more aquatic in the past?* Bentham e-book, 2012.

8. 'Human Evolution: Past, Present & Future', International Conference, London, May 8-10, 2013.

Chapter 2: Evolution revisited

1. Huxley, J., *Evolution: the modern synthesis,* Allen & Unwin, London, 1942.

2. Kruska, D. 'Mammalian domestication and its effect on the brain structure and behavior' in *Intelligence and evolutionary biology*: 211-250. Jerison, I. (Ed) Berlin, Eidelberg: Springer. 1988.

3. Kruska, D. 'The effect of domestication on brain size and composition in the mink', *J. Zool.*, London, 1996; 239: 645-61.

4. www.primalhealthresearch.com

5. Bucher, E. 'The return of Lamarck?' *Front. Gene* 2013. 4:10. doi: 10.3389/fgene.2013.00010.

Chapter 3: The future of the human oxytocin system

1. Wax et al. 'Maternal and newborn outcomes in planned home birth vs planned hospital births: a meta-analysis' *American Journal of Obstetrics & Gynecology*, 2010; 203 (3) DOI: 10.1016/j. ajog.2010.05.028.

2. Evers, A.C., Brouwers, H.A., Hukkelhoven, C.W., et al. 'Perinatal mortality and severe morbidity in low and high risk term pregnancies in the Netherlands: prospective cohort study' *BMJ* 2010; 341:c5639 doi: 10.1136/bmj.c5639 (Published 2 November 2010)

3. Birthplace in England Collaborative Group, 'Perinatal and maternal outcomes by planned place of birth for healthy women with low risk pregnancies: the Birthplace in England national prospective cohort study' *BMJ* 2011 Nov 23;343:d7400. doi: 10.1136/bmj.d7400.

4. Li, H-T., Ye, R., Achenbach, T., Ren, A., Pei, L., Zheng, X., Liu, J-M., 'Caesarean delivery on maternal request and childhood psychopathology: a retrospective cohort study in China', *BJOG* 2010; DOI: 10.1111/j.1471-0528.2010.02762.x.

5. Laughon, S.K., Branch, D.W., Beaver, J., Zhang, J., 'Changes in labor patterns over 50 years' *Am J Obstet Gynecol.* 2012 May;206(5):419. e1-9. Epub 2012 Mar 10.

6. http://www.oecd.org/dataoecd/30/56/43136964.pdf

7. Lawmann, E.O., Paik, A., Rosen, R.C., 'Sexual Dysfunction in the United States', *JAMA* 1999; 281: 537-544.

8. Konrath, S.H., O'Brien, E.H., Hsing, C. 'Changes in dispositional empathy in American college students over time: a meta-analysis' *Pers Soc Psychol Rev.* 2011 May;15(2):180-98. Epub 2010 Aug 5.

Chapter 4: A landmark in the evolution of brain size?

1. The contents of this chapter have already been presented in: Odent, M., *The Caesarean*, Free Association Books, London, 2004.

2. Odent, M., *Primal Health*, Century-Hutchinson ,1986 (first edition). Clairview, 2002 (second edition).

3. Odent, M., 'Prématurité et créativité' in *Les cahiers du nouveau-né* no. 6. Un enfant, prématurément, Le Vaguerese ed. Stock, Paris, 1983.

4. Odent, M., McMillan, L., Kimmel, T., 'Prenatal care and sea fish', *Eur J Obstet Gynecol.* 1996;68:49-51.

5. Meeson, L.F., 'The effects on birth outcomes of discussions in early pregnancy, emphasising the importance of eating fish', PhD thesis, University of Wolverhampton, July 31, 2007.

6. English, J. *Different doorway: adventures of a cesarean born,* Earth Heart, 1985

Chapter 5: 'Microbes Maketh Man'

1. Dubos, R. 'Staphylococci and infection immunity', *Am J Dis Child* 1966; 105: 643-45.

2. Samuli Rautava, S., Maria Carmen Collado, M.C., Seppo Salminen, S., Erika Isolauri, E. 'Probiotics Modulate Host-Microbe Interaction in the Placenta and Fetal Gut: A Randomized, Double-Blind, Placebo-Controlled Trial', *Neonatology* 2012;102:178-184 (DOI: 10.1159/000339182)

3. Heijtz, R.D., Wang, S., Anuar, F., et al. 'Normal gut microbiota modulates brain development and behavior', *Proc Natl Acad Sci* USA 2011 Feb 15;108(7):3047-52. Epub 2011 Jan 31.

4. Capone, K.A., Dowd, S.E., Stamatas, G.N., Nikolovski, J. 'Diversity of the human skin microbiome early in life', *J Invest Dermatol* 2011 Oct;131(10):2026-32. doi: 10.1038/jid.2011.168. Epub 2011 Jun 23,

5. Hulcr, J., Latimer, A.M., et al. 'A Jungle in There: Bacteria in Belly Buttons are Highly Diverse, but Predictable'. *PLoS One* 2012;7(11):e47712. doi: 10.1371/journal.pone.0047712. Epub 2012 Nov 7.

6. Virella, G., Silveira Nunes, M.A., and Tamagnini, G. 'Placental transfer of human IgG subclasses', *Clin Exp Immunol.* 1972 March; 10(3): 475–478

7. Aagaard, K., Riehle, K., Ma, J., et al. 'A metagenomic approach to characterization of the vaginal microbiome signature in pregnancy', *PLoS One* 2012;7(6):e36466. doi: 10.1371/journal.pone.0036466. Epub 2012 Jun 13.

8. Vaneechoutte, M. 'The similarities between the vaginal microflora and the gut microflora', presentation at the Mid-Pacific Conference

on Birth and Primal Health Research, Honolulu October 26-28.

9. Cederqvist, L.L., Ewool, L.C., Litwin, S.D. 'The effect of fetal age, birth weight, and sex on cord blood immunoglobulin values', *Am J Obstet Gynecol.* 1978 Jul 1;131(5):520-5

10. Garty, B.Z., Ludomirsky, A., Danon, Y., et al. 'Placental transfer of immunoglobulin G subclasses', *Clin Diagn Lab Immunol.* 1994 Nov;1(6):667-9.

11. Hashira, S., Okitsu-Negishi, S., Yoshino, K. 'Placental transfer of IgG subclasses in a Japanese population', *Pediatr Int.* 2000 Aug;42(4):337-42.

12. NICE's updated guideline on caesarean section is available at www. nice.org.uk/guidance/CG132.

13. Molloy, E.J., O'Neill, A.J., Grantham, J.J., Sheridan-Pereira, M., Fitzpatrick, J.M., Webb, D.W., Watson, R.W. 'Labor Promotes Neonatal Neutrophil Survival and Lipopolysaccharide Responsiveness', *Pediatr Res* 2004 May 5.

14. Gronlund, M.M., Nuutila, J., Pelto, L., Lilius, E.M., Isolauri, E., Salminen, S., Kero, P., Lehtonen, O.P. 'Mode of delivery directs the phagocyte functions of infants for the first 6 months of life', *Clin Exp Immunol* 1999; 116(3): 521-6.

15. Gronlund, M.M., Lehtonen, O.P., Eerola, E., Kero, P. 'Fecal microflora in healthy infants born by different methods of delivery: permanent changes in intestinal flora after cesarean delivery', *J Pediatr Gastroenterol Nutr* 1999; 28(1): 19-25.

16. van Nimwegen, F.A., Penders, J., Stobberingh, E.E., et al. 'Mode and place of delivery, gastrointestinal microbiota, and their influence on asthma and atopy', *J Allergy Clin Immunol.* 2011 Nov;128(5):948-55.e1-3. Epub 2011 Aug 27.

17. Turnbaugh, P.J., Ley, R.E., Mahowald, M.A., et al. 'An obesity-associated gut microbiome with increased capacity for energy harvest', *Nature* 21 December 2005; 444:1027-1031.

18. Huh, S.Y., Rifas-Shiman, S.L., Zera, C.A., et al. 'Delivery by caesarean section and risk of obesity in preschool age children: a prospective cohort study', *Arch Dis Child.* 2012 Jul;97(7):610-6. Epub 2012 May 23.

19. Larsen, N., Vogensen, F.K., Van der Berg, F.W.J., et al. 'Gut microbiota in human adults with type 2 diabetes differs from non-diabetic adults', *PloS One* February 5, 2010; 5(2): e9085.

20. Lif Holgerson, P., Harnevik, L., Hernell, O., et al. 'Mode of birth delivery affects oral microbiota in infants', *J Dent Res.* 2011 Oct;90(10):1183-8. Epub 2011 Aug 9.

21. Bik, E., 'Effect of mode of delivery on bacterial colonization of newborns', presentation at the Mid-Pacific Conference on Birth and

Primal Health Research, Honolulu October 26-28.

22. Cabrera-Rubio, R., Collado, M.C., Laitinen, K., et al. 'The human milk microbiome changes over lactation and is shaped by maternal weight and mode of delivery', *Am J Clin Nutr.* 2012 Sep;96(3):544-51. doi: 10.3945/ajcn.112.037382. Epub 2012 Jul 25.

23. Azad, M.B., Konya, T., Maugham, H., et al. 'Gut microbiota of healthy Canadian infants: profiles by mode of delivery and infant diet at 4 months', *CMAJ* February 11, 2013 cmaj.

24. Béchamp, A. *Les microzymas*, Paris, 1883. New edition 1990 by Centre International d'Etudes A. Béchamp.

Chapter 6: Should we criminalize planned vaginal birth?

1. Al-Mufti, R., McCarthy, A., Fisk, N.M., 'Survey of obstetricians' personal preference and discretionary practice', *Eur J Obstet Gynecol Reprod Biol* 1997; 73: 1-4.

2. Gabbe, S.G., Holzman, G.B. 'Obstetricians' choice of delivery', *Lancet* 2001; 357: 722.

3. Krebs, L., Langhoff-Roos, J. 'Elective cesarean delivery for term breech', *Obstet Gynecol* 2003; 101(4): 690-6

4. Harper, M.A., Byington, R.P., Espeland, M.A., et al. 'Pregnancy-Related Death and Health Care Services', *Obstet & Gynecol* 2003; 102(2): 273-278.

5. Liu, S., Liston, R.M., Joseph, K.S., et al. 'Maternal mortality and severe morbidity associated with low-risk planned cesarean delivery versus planned vaginal delivery at term', *CMAJ* 2007;176(4):455-60.

6. Hankins, G.D., Clark, S.M., Munn, M.B. 'Cesarean section on request at 39 weeks: impact on shoulder dystocia, fetal trauma, neonatal encephalopathy, and intrauterine fetal demise', *Semin Perinatol.* 2006 Oct;30(5):276-87.

Chapter 7: That is the question

1. Odent, M. *Primal Health*, Century Hutchinson, London 1986 (paperback 1987).

2. Kramer, M.S., Joseph, K.S. 'Enigma of fetal/infant-origins hypothesis', *Lancet* 1996; 348: 1254-55

3. Odent, M. *Primal Health*, 2nd edition, Clairview Books, Forest Row, (UK) 2002.

4. Krehbiel, D., Poindron, P., et al. 'Peridural anaesthesia disturbs maternal behaviour in primiporous and multiporous parturient ewes', *Physiology and behavior* 1987; 40: 463-72

5. Lundbland, E.G., Hodgen, G.D. 'Induction of maternal-infant bonding in rhesus and cynomolgus monkeys after caesarian delivery', *Lab. Anim. Sci* 1980; 30: 913.

6. Soto-Ramírez, N., Arshad, S., Holloway, J., et al. *Clinical Epigenetics* 2013, 5:1 doi:10.1186/1868-7083-5-1

7. Odent, M. 'Autism and Anorexia Nervosa: two facets of the same disease?' *Medical Hypotheses* 2010. doi:10.1016/j.mehy.2010.01.039

Chapter 8: Active management of human evolution

1. Odent, M. *The Functions of the Orgasms: the highways to transcendence*, Pinter and Martin, London, 2009.

2. De Catanzaeo, D. 'Suicide and self-damaging behavior: a sociobiological perspective', New York, Academic Press, 1981.

3. Voracek, M. 'Cross-national social ecology of intelligence and suicide prevalence: integration, refinement, and update of studies', *Percept Mot Skills* 2008 Apr;106(2):550-554.

4. Simon-Areces, J., Dietrich, M.O., Hermes, G., et al. 'UCP2 Induced by Natural Birth Regulates Neuronal Differentiation of the Hippocampus and Related Adult Behavior', *PLoS ONE*, 2012; 7 (8): e42911 DOI: 10.1371.

5. Odent, M. 'Making sense of rising caesarean section rates: reducing caesarean section rates should not be the primary objective', *BMJ* 2004 Nov 20;329(7476):1240.

6. Odent, M. 'If I were the baby. Questioning the widespread use of synthetic oxytocin', *Midwifery Today* 2010;94:22-3.

7. Gonzalez Mesa, E. 'Hard data about the side effects of synthetic oxytocin', presentation at the Mid-Pacific Conference on Birth and Primal Health Research, Honolulu, October 26, 2012.

8. Odent, M. *Childbirth in the age of plastics*, Pinter & Martin, London 2011.

9. Olza Fernández, I., Marín Gabriel, M., Malalana Martínez, A., et al. 'Newborn feeding behaviour depressed by intrapartum oxytocin: a pilot study', *Acta Paediatr*. 2012 Jul;101(7):749-54. doi: 10.1111/j.1651-2227.2012.02668.x. Epub 2012 Apr 4.

10. Bell, A.F., White-Traut, R., Rankin, K. 'Fetal exposure to synthetic oxytocin and the relationship with prefeeding cues within one hour postbirth', *Early Hum Dev*. 2012 Oct 16. pii: S0378-3782(12)00239-3. doi: 10.1016/j.earlhumdev.2012.09.017. [Epub ahead of print]

Chapter 9: Physiology v cultural conditioning

1. Odent, M. *Genese de l'homme écologique* Epi. Paris, 1979.

Chapter 10: Reasonable optimism

1. Engelmann, George J. *Labor Among Primitive Peoples*, J.H. Chambers & Co., St. Louis, 1884.

2. Odent, M. 'Colostrum and civilization', in Odent, M. *The Nature of*

Birth and Breastfeeding, Bergin & Garvey, 1992. 2nd ed 2003 (*Birth and Breastfeeding*, Clairview).

3. Odent, M. 'Neonatal tetanus', *Lancet* 2008;371:385-386

4. Fildes, V.A. *Breasts, bottles and babies. A history of infant feeding*, Edinburgh University Press, 1986.

5. Klopfer, M.S., Adams, D.K. 'Maternal imprinting in goats', proceedings of the National Academy of Sciences (USA) 1964;52:911-914.

6. Klaus, M.H., Kennell, J.H. 'Maternal-infant bonding', 1976. CV Mosby, St Louis.

7. De Chateau, P., Wiberg, B. 'Long-term effect on mother-infant behavior of extra contact during the first hour postpartum. I. First observations at 36 hours', *Acta Paediatrica Scand* 1977;66:137.

8. De Chateau, P., Wiberg, B. 'Long-term effect on mother-infant behavior of extra contact during the first hour postpartum. II. Follow-up at three months', *Acta Paediatrica Scand* 1977;66:145.

9. Schaller, J., Carlsson, S.G., Larsson, K. 'Effects of extended post-partum mother-child contact on the mother's behavior during nursing', *Infant Behavior and Development* 1979 (2):319-324

10. Terkel, J., Rosenblatt, J.S. 'Humoral factors underlying maternal behaviour at parturition: cross transfusion between freely moving rats', *J Comp Physiol Psychol* 1972;80: 365-371.

11. Siegel, H.I., Greenwald, M.S. 'Effects of mother-litter separation on later maternal responsiveness in the hamster', *Physiol Behav* 1978;21:147-149.

12. Siegel, H.I., Rosenblatt, J.S. 'Estrogen-induced maternal behaviour in hysterectomized-ovariectomized virgin rats', *Physiol Behav* 1975;465-471.

13. Jeliffe, D.B., Jeliffe, E.F.P. (Eds). 'The uniqueness of human milk', *Am J Clin Nutr* 1971;24:968-1009.

14. Jeliffe, D.B., Jeliffe, E.F.P. *Human milk in the modern world*, Oxford University Press, 1978.

15. McClelland, D.B., McGrath, J., Samson, R.R. 'Antimicrobial factors in human milk. Studies of concentration and transfer to the infant during the early stages of lactation', *Acta Paediatr Scand* Suppl. 1978;(271):1-20.

16. Odent, M. 'The early expression of the rooting reflex', proceedings of the fifth International Congress of Psychosomatic Obstetrics and Gynaecology, Rome 1977. London: Academic Press, 1977: 1117-19.

17. Odent, M. 'L'expression précoce du réflexe de fouissement', in *Les cahiers du nouveau-né* 1978; 1-2: 169-185 .

18. Virella, G., Silveira Nunes, M.A., Tamagnini, G. 'Placental transfer of human IgG subclasses', *Clin Exp Immunol.* 1972 Mar;10(3):475-8.

19. Pitcher-Wilmott, R.W., Hindocha, P., Wood, C.B. 'The placental transfer of IgG subclasses in human pregnancy', *Clin Exp Immunol.* 1980 Aug;41(2):303-8.

Chapter 11: Avenues for research

1. Uvnas Moberg, K. *The oxytocin factor*. Da Capo Press, Cambridge MA, 2003.
2. Pose, S.V., Cibils, L.A., Zuspan, F.P. 'Effect of l-epinephrine infusion on uterine contractility and cardiovascular system', *Am J Obstet Gynecol* 1962 84(3):297-306.
3. Wittlestone, W.G. 'The effect of adrenaline on the ejection response of the sow', *J Endocrin* 1954 10:167-172.
4. Lederman, R.P., McCann, D.S., Work, B., Huber, M.J. 'Endogenous plasma epinephrine and norepinephrine in last-trimester pregnancy and labor', *Am J Obstet Gynecol* 1977 129:5-8.
5. Lederman, R.P., Lederman, E., Work, B., McCann, D.S. 'The relationship of maternal anxiety, plasma catecholamines, and plasma cortisol to progress in labor', *Am J Obstet Gynecol* 1978 132(5):495-500.
6. Paterson, R., Seath, J., Taft, P., Wood, C. 'Maternal and foetal ketone concentrations in plasma and urine', *Lancet* 1967; ii: 862-5.
7. Mendiola, J., Grylack, L.J., Scanlon, J.W. 'Effects of intrapartum maternal glucose infusion on the normal fetus and newborn', *Anesth Analg.* 1982 Jan;61(1):32-5.
8. Lucas, A., Adrian, T.E., Aynsley-Green, A., Bloom, S.R. 'Iatrogenic hyperinsulinism at birth', *Lancet* 1980 Jan 19;1(8160):144-5.
9. Kenepp, N.B., Kumar, S., Shelley, W.C., Stanley, C.A., Gabbe, S.G., Gutsche, B.B. 'Fetal and neonatal hazards of maternal hydration with 5% dextrose before caesarean section', *Lancet* 1982 May 22;1(8282):1150-2.
10. Carmen, S. 'Neonatal hypoglycemia in response to maternal glucose infusion before delivery', *J Obstet Gynecol Neonatal Nurs.* 1986 Jul-Aug;15(4):319-23.

Chapter 12: Repressed common sense

1. Odent, M. 'Why labouring women don't need support', *Mothering* 1996;80:46-51.
2. McDonald, P., *The Oxford Dictionary of Medical Quotations*, p78, Oxford University Press, 2004.
3. Odent, M. 'Knitting Midwives for drugless childbirth?' *Midwifery Today* 2004; 71: 21-22.
4. Chunling Li, Weidong Wang, Sandra N. Summer, et al. 'Molecular Mechanisms of Antidiuretic Effect of Oxytocin', *J Am Soc Nephrol.*

2008 February; 19(2): 225–232.

5. Johansson, S., Lindow, S., Kapadia, H., Norman, M. 'Perinatal water intoxication due to excessive oral intake during labour', *Acta paediatr.* 2002;91(7):811-4.

6. Ophir, E., Solt, I., Odeh, M., Bornstein, J. 'Water intoxication - a dangerous condition in labor and delivery rooms', *Obstet Gynecol Surv.* 2007 Nov;62(11):731-8.

7. McCartney, M. 'Waterlogged?' *BMJ* 2011;343:d4280.

Chapter 13: The story is not finished

1. Nissen, E., Lilja, G., Widström, A.M., Uvnäs-Moberg, K. 'Elevation of oxytocin levels early post partum in women', *Acta Obstet Gynecol Scand.* 1995 Aug;74(7):530-3.

2. Lederman, R.P., Lederman, E., Work, B.A., McCann, D.S. 'Anxiety and epinephrine in multiparous women in labor: relationship to duration of labor and fetal heart rate pattern', *Am J Obstet Gynecol* 1985;153:870-78.

3. Nissen, E., Uvnas-Moberg, K., Svensson, K., Stock, S., Widstrom, A.M., Winberg, J. 'Different patterns of oxytocin, prolactin but not cortisol release during breastfeeding in women delivered by caesarean section or by the vaginal route', *Early Human Development* 1996; 45: 103-18.

4. Csontos, K., Rust, M., Hollt, V., et al. 'Elevated plasma beta-endorphin levels in pregnant women and their neonates', *Life Sci.*1979; 25: 835-44.

5. Akil, H., Watson, S.J., Barchas, J.D., Li, C.H. 'Beta-endorphin immunoreactivity in rat and human blood: Radioimmunoassay, comparative levels and physiological alterations', *Life Sci.* 1979; 24: 1659-66.

Chapter 14: Labour pain revisited

1. Csontos, K., Rust, M., Hollt, V., et al. 'Elevated plasma beta-endorphin levels in pregnant women and their neonates', *Life Sci.*1979; 25: 835-44.

2. Akil, H., Watson, S.J., Barchas, J.D., Li, C.H. 'Beta-endorphin immunoreactivity in rat and human blood: Radioimmunoassay, comparative levels and physiological alterations', *Life Sci.* 1979; 24: 1659-66.

3. Rivier, C., Vale, W., Ling, N., Brown, M., Guillemin, R. 'Stimulation in vivo of the secretion of prolactin and growth hormone by beta-endorphin', *Endocrinology* 1977; 100: 238-41.

4. Odent, M. 'La reflexotherapie lombaire. Efficacité dans le traitement de la colique néphrétique et en analgésie obstétricale', *La Nouvelle*

Presse Medicale 1975; 4 (3):188

5. Odent, M. 'Birth under water', *Lancet* 1983: pp1476-77.

6. Lieberman, E., Davidson, K., Lee-Parritz, A., Shearer E. 'Changes in fetal position during labor and their association with epidural analgesia', *Obstet Gynecol.* 2005 May;105(5 Pt 1):974-82.

7. Beilin, Y., Bodian, C., Weiser, J., et al. 'Effect of labor epidural analgesia with and without fentanyl on infant breast-feeding: a prospective, randomized, double-blind study', *Anesthesiology* 2005 Dec;103(6):1211-7.

8. Kurth, L., Haussmann, R. 'Perinatal Pitocin as an early ADHD biomarker: neurodevelopmental risk?' *J Atten Disord.* 2011 Jul;15(5):423-31. doi:10.1177/1087054710397800. Epub 2011 Apr 28.

9. Gonzalez Mesa, E. 'Hard data about the side effects of synthetic oxytocin', presentation at the Mid-Pacific Conference on Birth and Primal Health Research, Honolulu, October 26-28, 2012.

10. Dalman, C., Thomas, H.V., et al. 'Signs of asphyxia at birth and risk of schizophrenia. Population-based case-control study', *Br J Psychiatry* 2001 Nov;179:403-8.

11. Jacobson, B., Eklund, B., et al. 'Perinatal origin of adult self-destructive behavior', *Acta Psychiatr Scand.* 1987 Oct;76(4):364-71.

12. Holst, K., Andersen, E. 'Antenatal and perinatal conditions correlated to handicap among four-year-old children', *Am J Perinatol.* 1989 Apr;6(2):258-67.

13. Li, H-T., Ye, R., Achenbach, T., Ren, A., Pei, L., Zheng, X., Liu, J-M. 'Caesarean delivery on maternal request and childhood psychopathology: a retrospective cohort study in China', *BJOG* 2010; DOI: 10.1111/j.1471-0528.2010.02762.x.

14. Taylor, A., Fisk, N.M., Glover, V. 'Mode of delivery and subsequent stress response', *Lancet* 2000;355:120.

15. Cabrera-Rubio, R., Collado, M.C., Laitinen, K., et al. 'The human milk microbiome changes over lactation and is shaped by maternal weight and mode of delivery', *Am J Clin Nutr* 2012. Sep;96(3):544-51.

16. Azad, M.B., Konya, T., Maugham, H., et al. 'Gut microbiota of healthy Canadian infants: profiles by mode of delivery and infant diet at 4 months', *CMAJ* February 11, 2013.

Chapter 15: No paradigm shift without language shift

1. Newton, N., Foshee, D., Newton, M. 'Experimental inhibition of labor through environmental disturbance', *Obstet Gynecol* 1966;67:371-377.

2. Newton, N. 'The fetus ejection reflex revisited', *Birth* 1987;14(2):106-108.

REFERENCES

3. Odent, M. 'The fetus ejection reflex', *Birth* 1987;14(2):104-105.
4. Odent, M. 'New reasons and new ways to study birth physiology', *Int J Gynecol Obstet*. 2001;75 Suppl 1: S39-S455.
5. Odent, M. 'Fear of death during labour', *Journal of Reproductive and Infant Psychology* 2001;9:43-47.
6. Odent, M. 'Why laboring women don't need support', *Mothering* 1996; 80: 46-51.
7. Odent, M. 'The second stage as a disruption of the fetus ejection reflex', *Midwifery Today Int Midwife*, 2000 Autumn;(55):12.
8. Wen, S.W., Liu, S., Kramer, M.S., et al. 'Impact of prenatal glucose screening on the diagnosis of gestational diabetes and on pregnancy outcomes', *Am J Epidemiol* 2000; 152(11): 1009-14.
9. Brody, S.C., Harris, R., Lohr, K. 'Screening for gestational diabetes: a summary of the evidence for the U.S. Preventive Services Task Force', *Obstet Gynecol*. 2003 Feb;101(2):380-92.
10. Odent, M. 'Gestational diabetes and Health Promotion', *Lancet* 2009;374:684.
11. Bentley-Lewis, R. 'Gestational diabetes mellitus: an opportunity of a lifetime', *Lancet* 2009;373:1738-1740.
12. Gillman, M.W., Oakey, H., Baghurst, P.A. 'Effect of treatment of gestational diabetes mellitus on obesity in the next generation', *Diabetes Care* 2010 May;33(5):964-8. doi: 10.2337/dc09-1810. Epub 2010 Feb 11.
13. Flaxman, S.M., Sherman, P.W. 'Morning sickness: a mechanism for protecting mother and embryo', *Q Rev Biol*. 2000 Jun;75(2):113-48.
14. Weigel, R.M., Weigel, M.M. 'Nausea and vomiting of early pregnancy and pregnancy outcome. A meta-analytical review', *Br J Obstet Gynaecol*. 1989 Nov;96(11):1312-8.
15. Tierson, F.D., Olsen, C.L., Hook, E.B. 'Nausea and vomiting of pregnancy and association with pregnancy outcome', *Am J Obstet Gynecol*. 1986 Nov;155(5):1017-22.
16. Buckwalter, J.G., Simpson, S.W. 'Psychological factors in the etiology and treatment of severe nausea and vomiting in pregnancy', *Am J Obstet Gynecol*. 2002 May;186(5 Suppl Understanding):S210-4.
17. Czeizel, A.E., Sarhozi, A., Wyszynski, D.F. 'Protective effect of hyperemesis gravidarum for nonsyndromic oral clefts', *Obstet Gynecol* 2003 Apr;101(4):737-44.
18. Akre, O., Boyd, H.A., Ahlgren, M., 'Maternal and gestational risk factors for hypospadias', *Environ Health Perspect*. 2008 Aug;116(8):1071-6. doi: 10.1289/ehp.10791.
19. Eliakim, R., Abulafia, O., Sherer, D.M. 'Hyperemesis gravidarum: a current review', *Am J Perinatol*. 2000;17(4):207-18.
20. de Aloysio, D., Penacchioni, P. 'Morning sickness control in

early pregnancy by Neiguan point acupressure', *Obstet Gynecol* 1992;80(5):852-4.

21. Bolin, M., Akerud, H., Cnattingius, S., Stephansson, O., Wikström, A. 'Hyperemesis gravidarum and risks of placental dysfunction disorders: a population-based cohort study', *BJOG* 2013 Jan 30. doi: 10.1111/1471-0528.12132. [Epub ahead of print]

Chapter 16: Love as an evolutionary handicap

1. Eaton, S.B., Shostak, M., Konner, M. *The paleolithic prescription,* Harper and Row, New York, 1988.
2. Hallet, J.P., *Pygmy Kitabu* Random House, NY 1973.
3. Schiefenhovel, W. *Childbirth among the Eipos, New Guinea*, film presented at the Congress of Ethnomedicine, 1978, Gottingen, Germany.
4. Everett, D., *Don't sleep, there are snakes* Profile Books, 2008.
5. Odent, M. *The Scientification of Love*, Free Association Books, London, 1999.
6. Mead, M. *Sex & temperament in three primitive societies,* William Morrow and Co., NY, 1935.
7. Engelmann, G.J. *Labor Among Primitive Peoples*, J.H. Chambers & Co., St Louis, 1884.
8. Sobonfu Somé, *Welcoming Spirit Home: Ancient African Teachings to celebrate children and community* Novato, CA: New world library, 1999.

Chapter 17: Reasonable pessimism

1. Odent, M., *Primal Health*, Century Hutchinson, London, 1986.
2. Sun, T., Patoine, C., Abu-Khalil, A., et al. 'Early asymmetry of gene transcription in embryonic human left and right cerebral cortex', *Science* 2005 Jun 17;308(5729):1794-8. Epub 2005 May 12.
3. Gupta, R.K., Hasan, K.M., Trivedi, R., et al. 'Diffusion tensor imaging of the developing human cerebrum', *J Neurosci Res.* 2005 Jul 15;81(2):172-8.
4. Lenzi, D., Trentini, C., Pantano, P. 'Neural basis of maternal communication and emotional expression processing during infant preverbal stage', *Cereb Cortex.* 2009 May;19(5):1124-33. doi: 10.1093/cercor/bhn153. Epub 2008 Sep 11.
5. Fessenden, M. 'Students with autism gravitate toward STEM majors', *Nature* doi:10.1038/nature.2013.12367.
6. Hultman, C., Sparen, P., Cnattingius, S. 'Perinatal risk factors for infantile autism', *Epidemiology* 2002; 13: 417-23.
7. Glemma, E.J., Bower, C., Petterson, B., et al. 'Perinatal factors and the development of autism', *Arch Gen Psychiatry* 2004; 61: 618-27.

8. Hattori, R., et al. 'Autistic and developmental disorders after general anaesthetic delivery', *Lancet* 1991; 337 : 1357-1358 (letter)

9. Stein, D., Weizman, A., Ring, A., Barak, Y. 'Obstetric complications in individuals diagnosed with autism and in healthy controls', *Compr Psychiatry* 2006 Jan-Feb;47(1):69-75.

10. Modahl, C., Green, L., et al. 'Plasma oxytocin levels in autistic children', *Biol Psychiatry* 1998; 43 (4): 270-7.

11. Green, L., Fein, D., et al. 'Oxytocin and autistic disorder: alterations in peptides forms', *Biol Psychiatry* 2001; 50 (8): 609-13.

12. Hambrook, D., Tchanturia, K., Schmidt, U., Russell, T., Treasure, J. 'Empathy, systemizing, and autistic traits in anorexia nervosa: A pilot study', *Br J Clin Psychol.* 2008 Sep;47(Pt 3):335-9. Epub 2008 Jan 21.

13. Southgate, L., Tchanturia, K., Treasure, J. 'Information processing bias in anorexia nervosa', *Psychiatry Res.* 2008 Jun 23. (Epub ahead of print)

14. Wentz, E., Lacey, J.H., Waller, G., Råstam, M., Turk, J., Gillberg, C. 'Childhood onset neuropsychiatric disorders in adult eating disorder patients. A pilot study', *Eur Child Adolesc Psychiatry* 2005 Dec;14(8):431-7

15. Cnattingius, S., Hultman, C.M., Dahl, M., Sparen, P. 'Very preterm birth, birth trauma and the risk of anorexia nervosa among girls', *Arch Gen Psychiatry* 1999 56: 634-38.

16. Favaro, A., Tenconi, E., Santonastaso, P. 'Perinatal factors and the risk of developing anorexia nervosa and bulimia nervosa', *Arch Gen Psychiatry* 2006 63(1):82-8.

17. Odent, M., 'Autism and anorexia nervosa. Two facets of the same disease?' *Med Hypotheses* 2010 . doi:10.1016/j.mehy.2010.01.039

18. Demitrack, M.A., Lesem, M.D., Listwak, S.J., et al. 'CSF oxytocin in anorexia nervosa and bulimia nervosa: clinical and pathophysiologic considerations', *Am J Psychiatry* 1990 Jul;147(7):882-86.

19. Gunter, H.L., Ghaziuddin, M., Ellis, H.D. 'Asperger symdrom: tests of right hemisphere functioning and interhemispheric communication', *Journal of Autism and Developmental Disorders* 2002; 32(4):263-81.

20. Waiter, G.D., Williams, J.H., Murray, A.D., et al. 'Structural white matter deficits in high-functioning individuals with autistic spectrum disorders: a voxel-based investigation', *NeuroImage* 2005; 24(2): 455-61.

21. Maxwell, J.K., Tucker, D.M., Townes, B.D. 'Asymmetric cognitive function in anorexia nervosa', *International Journal of Neuroscience* 1984; 24(1): 37-44.

22. Uher, R., Murphy, T., Friederich, H.C., et al. 'Functional

neuroanatomy of body shape perception in healthy and eating-disordered women', *Biol Psychiatry* 2005;58(12):990-97.

23. Grunwald, M., Weiss, T., Assmann, B., et al. 'Stable asymmetric theta power in patients with anorexia nervosa during haptic perception even after weight gain: a longitudinal study', *Journal of Experimental Neuropsychology* 2004; 26(5):608-20.

24. Uher, R., Treasure, J. 'Brain lesions and eating disorders', *Journal of Neurology, Neurosurgery, and Psychiatry* 2005; 76(6): 852-7.

25. Oberman, L.M., Hubbard, E.M., McCleery, J.P., Altschuler, E.L., Ramachandran, V.S., Pineda, J.A., 'EEG evidence for mirror neuron dysfunction in autism spectral disorders', *Brain Res Cogn.*; 24(2):190-8, 2005-06.

26. Dapretto, M., 'Understanding emotions in others: mirror neuron dysfunction in children with autism spectrum disorders', *Nature Neuroscience*, Vol. 9, No. 1, pp. 28-30, 2006-01.

27. Culbert, K.M., Breedlove, S.M., Burt, S.A., Klump, K.L., 'Prenatal hormone exposure and risk for eating disorders: a comparison of opposite-sex and same-sex twins', *Arch Gen Psychiatry*, 2008 Mar;65(3):329-36.

28. Devlin, B., Bacanu, S.A., Klump, K.L. et al. 'Linkage analysis of anorexia nervosa incorporating behavioral covariates', *Hum Mol Genet.* 2002 Mar 15;11(6):689-96

29. Konrath, S.H., O'Brien, E.H., Hsing, C. 'Changes in dispositional empathy in American college students over time: a meta-analysis', *Pers Soc Psychol Rev.* 2011 May;15(2):180-98. Epub 2010 Aug 5.

30. Burke, H.M., Davis, M.C., Otte, C., Mohr, D.C. 'Depression and cortisol responses to psychological stress: a meta-analysis', *Psychoneuroendocrinology* 2005 Oct;30(9):846-56.

31. Schore, A.N. 'The effect of relational trauma on right brain development, affect regulation, and infant mental health', *Infant Mental Health Journal* 2001;22: 201-269.

32. Veijola, J., Mäki, P., Joukamaa, M., et al. 'Parental separation at birth and depression in adulthood: a long-term follow-up of the Finnish Christmas Seal Home Children', *Psychol Med.* 2004 Feb;34(2):357-62.

33. *Der Zauberlehrling*, Goethe, 1797.

34. Cortese, S., Angriman, M., Maffeis, C., et al. 'Attention-deficit/hyperactivity disorder (ADHD) and obesity: a systematic review of the literature', *Crit Rev Food Sci Nutr.* 2008 Jun;48(6):524-37.

Chapter 19: *Homo sapiens* and the virosphere

1. Odent, M., *Primal Health*, Century Hutchinson, London, 1986.
2. Cohen, N., Moynihan, J.A., Ader, R. 'Pavlovian conditioning of the

immune system', *Int Arch Allergy Immunol.* 1994 Oct;105(2):101-6.

3. Gardner, S.G., Bingley, P.J., et al. 'Rising incidence of insulin dependent diabetes in children aged under 5 years in the Oxford region: time trend analysis', *BMJ* 1997; 315:713-7.

4. McKinney, P.A., Parslow, R., Gurney, K.A., Law, G.R., Bodansky, H.J., Williams, R. 'Perinatal and neonatal determinants of childhood type 1 diabetes. A case-control study in Yorkshire, U.K.', *Diabetes Care* 1999 Jun;22(6):928-32.

5. Cardwell, C.R., Stene, L.C., Joner, G., et al. 'Caesarean section is associated with an increased risk of childhood-onset type 1 diabetes mellitus: a meta-analysis of observational studies', *Diabetologia* 2008 May;51(5):726-35. Epub 2008 Feb 22.

6. Pistiner, M., Gold, D.R., et al. 'Birth by cesarean section, allergic rhinitis, and allergic sensitization among children with a parental history of atopy', *J Allergy Clin Immunol.* 2008 Aug;122(2):274-9. Epub 2008 Jun 20.

7. Bager, P., Wohlfahrt, J., Westergaard, T., 'Caesarean delivery and risk of atopy and allergic disease: meta-analyses', *Clin Exp Allergy* 2008. Apr;38(4):634-42.

8. Kuitunen, M., Kukkonen, K., Juntunen-Backman, K., et al. 'Probiotics prevent IgE-associated allergy until age 5 years in cesarean-delivered children but not in the total cohort', *J Allergy Clin Immunol.* 2009 Feb;123(2):335-41. doi: 10.1016/j.jaci.2008.11.019. Epub 2009 Jan 8.

9. Xu, B., Pekkanen, J., Hartikainen, A.L., Järvelin, M.R., 'Caesarean section and risk of asthma and allergy in adulthood', *J Allergy Clin Immunol.* 2001 Apr;107(4):732-3.

10. Roduit, C., Scholtens, S., de Jongste, J.C., et al. 'Asthma at 8 years of age in children born by caesarean section', *Thorax* 2009 Feb;64(2):107-13. doi: 10.1136/thx.2008.100875. Epub 2008 Dec 3.

11. Stensballe, L.G., Simonsen, J., Bisgaard, H., 'Use of Antibiotics during Pregnancy Increases the Risk of Asthma in Early Childhood', *J Pediatr.* 2012 Nov 6. pii: S0022-3476(12)01141-9. doi: 10.1016/j. jpeds.2012.09.049.

12. Johnson, N.P., Mueller, J. 'Updating the accounts: global mortality of the 1918-1920 'Spanish' influenza pandemic', *Bull Hist Med.* 2002 Spring;76(1):105-15.

13. Jones, N. 'Planetary disasters: it could happen one night', *Nature* News Feature. January 08, 2013.

14. Lovelock, James *Gaia: A new look at Life on Earth*, Oxford Paperbacks, 2000.

Chapter 20: Cultural blindness

1. Odent, M., 'Between circular and cul-de-sac epidemiology', *Lancet* 2000; 355 (April 15): 1371.

2. Jacobson, B., Nyberg, K., Gronbladh, L., et al. 'Opiate addiction in adult offspring through possible imprinting after obstetric treatment', *BMJ* 1990;301 (6760): 1067-70.

3. Tinbergen, N., Tinbergen, E.A., *Autistic children*, The Tinbergen Trust, 1983.

4. Hattori, R., Desimaru, M., Nagayama, I., Inoue, K. 'Autistic and developmental disorders after general anaesthetic delivery', *Lancet* 1991; 337:1357-58.

5. Glemma, E.J., Bower, C., Petterson, B., et al. 'Perinatal factors and the development of autism', *Arch Gen Psychiatry* 2004; 61: 618-27.

6. Hultman, C., Sparen, P., Cnattingius, S., 'Perinatal risk factors for infantile autism', *Epidemiology* 2002; 13: 417-23.

7. Stein, D., Weizman, A., Ring, A., Barak, Y. 'Obstetric complications in individuals diagnosed with autism and in healthy controls', *Compr Psychiatry* 2006 Jan-Feb;47(1):69-75.

8. Taylor, B., Miller, E., et al. 'Autism and measles, mumps, and rubella vaccine: no epidemiological evidence for a causal association', *Lancet* 1999; 353: 2026-9.

9. Kaye, J.A., Melero-Montes, M., Jick, H. 'Mumps, measles, and rubella vaccine and the incidence of autism recorded by general practitioners: a time trend analysis', *BMJ* 2001; 322: 460-3.

10. Dales, L., Hammer, S.J., Smith, N.J., 'Time trends in autism and in MMR immunization coverage in California', *JAMA* 2001; 285 (9): 1183-5.

11. Madsen, K.M., Hviid, A., et al. 'A population-based study of measles, mumps, and rubella vaccination and autism', *N Engl J Med* 2002; 347(19): 1474-5.

12. Hviid, A., Stellfeld, M., Wohlfahrt, J., Melbye, M., 'Association between thimerosal-containing vaccine and autism,' *JAMA* 2003 Oct 1;290(13):1763-6.

13. Levy, S.E., Mandell, D.S., Schultz, R.T., 'Autism', *Lancet* 2009 Nov 7;374(9701):1627-38. doi: 10.1016/S0140-6736(09)61376-3. Epub 2009 Oct 12.

14. Hughes, V. 'Epidemiology: Complex disorder', *Nature* 491; S2–S3. 01 November 2012. doi:10.1038/491S2a.

15. Montessori, M. *The Secret of Childhood*, Fides Publishers, 1966.

16. Rank, O., *The Trauma of Birth*, Martino Publishing, 2010.

17. Andreasen, N., *The Creating Brain*, Dana Press, 2005.

18. Horrobin, D., *The Madness of Adam and Eve*, Bantam Press, 2001.

19. Reich, W., *The Murder of Christ*, Farrar, Straus and Giroux, NY 1971.

REFERENCES

20. Leboyer, F., *Birth without Violence*, Fontana, 1977.
21. Kennell, J., Klaus, M., et al. 'Continuous emotional support during labor in an US hospital', *JAMA* 1991; 265: 2197-2201.

INDEX

INDEX